Progress in Mathematics
Volume 151

Christoph Hummel

Gromov's Compactness Theorem for Pseudo-holomorphic Curves

Birkhäuser Verlag
Basel · Boston · Berlin

Author:

Christoph Hummel
Institut für Mathematik   and   Department of Mathematics
Universität Zürich   University of Pennsylvania
Winterthurer Str. 190   209 South 33rd Street
8057 Zürich   Philadelphia, PA 19104–6395
Switzerland   USA

1991 Mathematics Subject Classification 53Cxx, 30Fxx

A CIP catalogue record for this book is available from the Library of Congress, Washington D.C., USA

Deutsche Bibliothek Cataloging-in-Publication Data

Hummel, Christoph:
Gromov's compactness theorem for pseudo-holomorphic curves /
Christoph Hummel. – Basel ; Boston ; Berlin : Birkhäuser, 1997
   (Progress in mathematics ; Vol. 151)
   ISBN 3-7643-5735-5 (Basel ...)
   ISBN 0-8176-5735-5 (Boston)

© 1997 Birkhäuser Verlag, P.O. Box 133, CH-4010 Basel, Switzerland
Printed on acid-free paper produced of chlorine-free pulp. TCF ∞
Printed in Germany
ISBN 3-7643-5735-5
ISBN 0-8176-5735-5

9 8 7 6 5 4 3 2 1

Für meine Eltern

Contents

Introduction

The present book is a revised, extended and translated version of my diploma thesis (Diplomarbeit) completed in December 1992 at the Albert-Ludwigs-Universität Freiburg im Breisgau under the guidance of Viktor Schroeder.

Overview

Pseudo-holomorphic curves are smooth maps from Riemann surfaces to almost complex manifolds with complex linear differential. They were introduced into symplectic geometry by M. Gromov in 1985. Our aim is to present a detailed proof of his compactness theorem for closed pseudo-holomorphic curves. The approach here follows Gromov's original proof [Gr], P. Pansu's notes [Pa1] and the articles of M.-P. Muller [Mu] and Pansu [Pa2]. Other approaches are due to T. Parker and J. Wolfson [PW], J.-C. Sikorav [Si] and Rugang Ye [Ye].

Gromov generalized well-known properties of holomorphic maps and in this process gave new geometric proofs of them. Using these properties together with properties of hyperbolic surfaces, he proved the compactness theorem. In addition to geometric properties of pseudo-holomorphic curves, we also include those basic facts about hyperbolic surfaces which are required for the understanding of the proof.

A complex structure on a real vector space V of even dimension is an endomorphism J_0 where J_0^2 is minus the identity. Such a structure turns V into a complex vector space. An almost complex manifold is a manifold M of even dimension together with an almost complex structure J, which is a complex structure on each tangent space of M depending smoothly on the base point. Riemann surfaces in particular are almost complex manifolds. Symplectic manifolds are important examples of manifolds which carry almost complex structures. A pseudo-holomorphic or J-holomorphic curve is a smooth map from a Riemann surface to an almost complex manifold such that its differential at each point is complex linear. In [Gr] Gromov stated existence results for J-holomorphic curves in symplectic manifolds. Since then J-holomorphic curves have become an important tool in symplectic geometry (see for instance [AL] and [ABK]).

In order to establish compactness properties of closed pseudo-holomorphic curves in almost complex manifolds the notion of pseudo-holomorphic curve is extended slightly to encompass so-called cusp curves. To illustrate this, consider the following

Example. Let $S^2 = \mathbb{C} \cup \{\infty\}$ denote the Riemann sphere. Then $f_n \colon S^2 \to S^2 \times S^2$ with $f_n(z) := (z, 1/(n^2 z))$, $n \geq 1$, is a sequence of holomorphic curves in $S^2 \times S^2$. Each f_n is an embedding. As $n \to \infty$ the images $f_n(S^2)$ converge to $S^2 \times \{0\} \cup \{0\} \times S^2$. In the limit the simple closed loops $f_n(\{\, |z| = 1/n \,\}) \subset S^2 \times S^2$ collapse to the point $(0,0)$ at which the two spheres $S^2 \times \{0\}$ and $\{0\} \times S^2$ are glued together.

Gromov's compactness theorem states that a sequence of closed J-holomorphic curves of the same genus and of uniformly bounded area in a compact, almost complex manifold has a subsequence converging to a cusp curve. In rough terms, this means that, after the removal of finitely many simple closed loops in each domain suitable reparametrizations converge uniformly and all derivatives converge locally uniformly. Moreover, in the image the removed loops collapse to points and the areas of the J-holomorphic curves converge. The limit cusp curve is a finite union of J-holomorphic curves, glued together at those points which are limits of collapsing closed loops.

Guide for the reader

In Chapter I we introduce basic notions such as Riemannian manifolds, almost complex and symplectic manifolds and J-holomorphic maps, and fix some of our notation. Sections addressing basic facts from 2-dimensional hyperbolic geometry and conformal annuli are also included.

Chapter II deals with Gromov's Schwarz lemma for J-holomorphic curves which gives uniform gradient estimates for certain families of J-holomorphic maps. It is a generalization of the classical Schwarz lemma for holomorphic maps. Essential for the proof are area estimates given by isoperimetric inequalities and the monotonicity lemma for pseudo-holomorphic curves with boundary which we also prove.

Chapter III shows how to estimate higher order derivatives. The basic idea is that one can force the differentials of pseudo-holomorphic maps — called the 1-jets — to be themselves pseudo-holomorphic maps. Then the Gromov-Schwarz lemma applies to 1-jets and in this way one can estimate second order derivatives of pseudo-holomorphic maps. Successively passing to 1-jets one can estimate higher order derivatives. Using this idea, Gromov generalized well-known properties of holomorphic maps, namely Riemann's theorem on the removal of singularities and a theorem of Weierstraß. The latter states that the derivatives of a locally uniformly convergent sequence of holomorphic maps also converge locally uniformly. This also holds for pseudo-holomorphic maps.

Chapter IV is concerned with the classical theory of hyperbolic surfaces as far as it is needed for the proof of the compactness theorem. Consider a closed connected and orientable surface S and let F be a finite subset containing more than two points. For each complex structure on S there is a unique complete Riemannian metric on $S \setminus F$, the so-called Poincaré metric, with constant curvature -1 and finite area such that the complex structure in each tangent space is a rotation by $\pi/2$. In particular, the Poincaré metric is a complete hyperbolic metric. A sequence (h_n) of complete hyperbolic metrics on $S \setminus F$ not having a convergent subsequence in the Riemann space, i.e., the space of all complete hyperbolic metrics on $S \setminus F$ modulo orientation preserving diffeomorphisms, is of the following type. The infimum of the lengths of simple closed geodesics in $(S \setminus F, h_n)$ converges to zero as $n \to \infty$. However, after the removal of finitely many simple closed geodesics for each metric, whose lengths converge to zero, the metrics on the resulting non-complete components converge to complete metrics. Thus the limit is a disjoint union of complete hyperbolic surfaces. This degeneration can be described precisely with the help of the pairs of pants decomposition.

In Chapter V, the compactness theorem is stated and proved. The structure of Gromov's proof is as follows. Consider a sequence (f_n) of closed J-holomorphic curves with diffeomorphic domains in a compact almost complex manifold. Choose a Hermitian metric on this manifold, that is a Riemannian metric invariant under the almost complex structure, and assume that the area of the J-holomorphic curves is uniformly bounded. Using the Gromov-Schwarz lemma one can show that, after removing a suitable, uniformly bounded number of points in the corresponding domains, the J-holomorphic curves are uniformly Lipschitz with respect to the corresponding Poincaré metrics. After passing to a subsequence the number of points removed can be assumed to be the same for each domain. The description of degenerating sequences of complete hyperbolic metrics on a fixed closed surface with finitely many points removed yields a uniform limit of some subsequence of (f_n), after a suitable reparametrization and removal of finitely many disjoint simple closed loops in the domain. Then the J-holomorphic curves are parametrized over the same surface with the same points removed but with different metrics. Moreover, the metrics converge outside the removed loops. However, in each metric the loops are simple closed geodesics with lengths converging to zero. And, since the maps are uniformly Lipschitz, their images collapse to points. The generalized Weierstraß theorem implies the locally uniform convergence of the derivatives and convergence of the areas. In particular, the limit is again J-holomorphic. It is also defined at those points removed at the beginning by the theorem on the removal of singularities. Convergence in a neighbourhood of these points follows from the monotonicity lemma.

Chapter VI describes an application of pseudo-holomorphic curves in symplectic geometry, namely Gromov's famous squeezing theorem which states a certain rigidity property of symplectic maps. We present Gromov's original proof of this theorem which contains a nice application of the monotonicity lemma and an application of the compactness theorem for pseudo-holomorphic curves. Since pseudo-holomorphic curves are solutions of a certain elliptic partial differential equation, a generalized Cauchy-Riemann equation, analytic tools are necessary in order to work with these curves. The previous chapters are mainly geometric and it is beyond the scope of this book to introduce the necessary analytic methods. However, we try to illustrate some of these methods with a particular existence result for pseudo-holomorphic curves which is used to show the squeezing theorem. In addition, we include a separate list of references about pseudo-holomorphic curves at the end of the book in order to give the reader ideas for further studies.

In Appendix A a short proof of the classical isoperimetric inequality for surfaces in Euclidean space is given. It plays an important role in the proof of the compactness theorem. Roughly speaking, it states that any smooth loop of length l in some Euclidean space can be filled with the image of some smooth map from the disc such that the area of the image is less or equal to the area of the standard Euclidean disc with perimeter l.

Appendix B introduces briefly the notion of C^k-convergence for sequences of maps between two manifolds where k is some non-negative integer or ∞.

In the present book, for reasons of convenience, we work in the C^∞-category. The proofs presented here also hold under weaker differentiability assumptions, for example some suitable C^k, as one can see without difficulties. There is also a compactness theorem for J-holomorphic curves with boundary which we shall not treat.

Prerequisites

We assume that the reader has had a first course on differentiable manifolds and is familiar with the notion of a vector bundle. For the latter, an understanding of §§2 and 3 of [MiS] is sufficient. The reader should also have some knowledge of covering spaces and [Ma] is a good reference on this subject. Some familiarity with elementary complex analysis and with the notion of a Riemannian manifold would also be helpful.

Acknowledgements

I am very much indebted to my teachers Prof. Victor Bangert and Prof. Viktor Schroeder. In their seminars and in many discussions with them, I acquired much of the material for this book and I profited greatly from their advice. I am very grateful to Prof. Helmut Hofer for his enormous encouragement and support in publishing these notes and for many valuable suggestions. Various friends and colleagues read parts of the manuscript and, with their comments and proposals, they helped a lot to improve this exposition. Among them, I wish to thank in particular Bernd Ammann, Prof. Enrico Leuzinger, Christof Luchsinger, Alexander McNeil, Felix Schlenk, Alexander Schmitt and Michel Walz. My special thanks go to my friend Harald Alferi for his immeasurable help and his criticism. He also designed the layout which he provided in his LaTeX-style *haRry*. The final part of this book was completed while I was enjoying a visit at the Department of Mathematics at the University of Pennsylvania and supported by the Swiss National Science Foundation.

Philadelphia, March 1, 1997

                                                          Christoph Hummel

Chapter I

Preliminaries

In this chapter we introduce basic objects and fix some of our notation and terminology.

1. Riemannian manifolds

We need only some elementary properties of Riemannian manifolds which we summarize here. For a brief introduction and outlines of proofs we refer to [Lf], especially Section 1. The reader will find additional material in several textbooks on Riemannian geometry, for instance [Ch], [dC], [GHL] or [Jo].

By *manifold* we always mean, unless otherwise stated, a smooth finite-dimensional manifold without boundary which has a countable basis of topology. The word *smooth* is always used as a synonym for C^∞-*differentiable*. For a differentiable map $f: M \to N$ between two manifolds M and N we denote by $Tf: TM \to TN$ its differential from the tangent bundle TM of M to the tangent bundle TN of N. Restricted to a tangent space T_pM to M at $p \in M$, it is a linear map $T_p f: T_pM \to T_{f(p)}N$.

A *Riemannian manifold* (M, g) is a manifold M together with a *Riemannian metric* g on M, that is a Euclidean scalar product g_p on each tangent space T_pM, $p \in M$, depending smoothly on p. Depending smoothly means that, given any two vector fields X and Y on M the function $g(X, Y): p \mapsto g_p(X(p), Y(p)) =: g(X(p), Y(p))$ is smooth. In local coordinates $x^1, \ldots, x^m$ a Riemannian metric is given by

$$g\left(\frac{\partial}{\partial x^\alpha}, \frac{\partial}{\partial x^\beta}\right) := g_{\alpha\beta},$$

where $g_{\alpha\beta}$, $\alpha, \beta = 1, \ldots m$, are smooth functions on the domain of the coordinates such that at any point the matrix $(g_{\alpha\beta})$ is symmetric and positive definite. Using a partition of unity argument, one can show the existence of a Riemannian metric on any manifold. Let $c: [a, b] \to (M, g)$ be a smooth path in a Riemannian manifold (M, g). Its *length* $\ell(c)$ is defined by

$$\ell(c) := \int_a^b \|\dot{c}(t)\| \, dt,$$

where $\dot{c}(t) = (dc/dt)(t) \in T_{c(t)}M$ and $\|v\| := \sqrt{g(v, v)}$ for any $v \in TM$. The length $\ell(c)$ is independent of the parametrization of c. The function $d: M \times M \to \mathbb{R}$ given by

$$d(p, q) := \inf\left\{ \ell(c) \mid c: [0, 1] \to M \text{ a smooth path connecting } p \text{ to } q \right\}$$

is well-defined if the Riemannian manifold (M, g) is connected. It is in fact a distance function on (M, g) inducing the topology of M and called *the distance function* of

(M, g). For $r > 0$ and $p \in M$ we denote by $B_r(p)$ and $\overline{B}_r(p)$ the *open* and *closed* r-ball in (M, g) around p, respectively. Of course, the previous definitions also make sense for manifolds with boundary.

Now consider a path $c: I \to (M, g)$ in a Riemannian manifold (M, g) defined on some interval $I \subset \mathbb{R}$. We call c a *geodesic segment* if it has constant speed $u \geq 0$ and is locally distance minimizing, i.e., for any $t_0 \in I$ there is a neighbourhood I' of t_0 in I such that $d(c(t), c(s)) = u \cdot |s - t|$ for all $s, t \in I'$. We call $c: I \to (M, g)$ a *geodesic* if it is maximal in the sense that for any geodesic segment $\tilde{c}: \tilde{I} \to M$ with $\tilde{I} \supset I$ and $\tilde{c}|_I = c$ we have $\tilde{I} = I$.

In local coordinates of M, geodesics are solutions of an ordinary differential equation of second order. This yields that, for any $v \in TM$ there is a unique geodesic $c_v: I_v \to M$ in (M, g) with $0 \in I_v \subset \mathbb{R}$ and $\dot{c}_v(0) = v$. Observe that $\|v\|$ is the speed of c_v. The set $\mathscr{D} := \{ v \in TM \mid 1 \in I_v \}$ is an open neighbourhood of the zero section of TM. The *exponential map* of (M, g) is the smooth map

$$\exp: \mathscr{D} \to M, \; v \mapsto c_v(1).$$

The *exponential map at* $p \in M$ is the map $\exp_p := \exp|_{\mathscr{D} \cap T_p M}$. By definition its differential $T_0 \exp_p: T_0 T_p M \to T_p M$ at $0 = 0_p \in T_p M$ is the identity after identifying $T_0 T_p M$ with $T_p M$ canonically. Hence $\exp_p$ is a local diffeomorphism near 0. The *injectivity radius of* (M, g) *at* $p \in M$ is

$$\mathrm{injrad}(M, g, p) := \sup \Big\{ r > 0 \mid \exp_p |_{B_r(0)}: B_r(0) \to \exp_p(B_r(0))$$
$$\text{is defined and is a diffeomorphism} \Big\}$$

where $B_r(0) \subset T_p M$ is the open r-ball in $T_p M$ around 0 with respect to g_p. Since $\exp_p$ is a local diffeomorphism near 0, the injectivity radius at $p \in M$ is greater than zero. Moreover, $\exp_p(B_r(0)) = B_r(p)$ for any $r \leq \mathrm{injrad}(M, g, p)$. The *injectivity radius of* (M, g) is

$$\mathrm{injrad}(M, g) := \inf_{p \in M} \mathrm{injrad}(M, g, p).$$

A Riemannian metric g on M is called *complete* if the induced distance function on each connected component is complete. In particular, compact Riemannian manifolds are complete. By the theorem of Hopf-Rinow, a Riemannian manifold is complete if and only if each geodesic is defined on $\mathbb{R}$. In a connected, complete Riemannian manifold any two given points can be joined by a geodesic segment of length equal to the distance of the two points.

Lemma 1.1. *If (M, g) is a compact Riemannian manifold then* $\mathrm{injrad}(M, g) > 0$.

Proof. Let $\pi: TM \to M$ be the projection onto the base points. Since $\exp_p$ is a local diffeomorphism near $0_p \in T_p M$, the map $E: TM \to M \times M, \; v \mapsto (\pi(v), \exp(v))$ is a local diffeomorphism near 0_p. For any $p \in M$ choose an open neighbourhood V_p of 0_p in TM such that $E|_{V_p}$ is a diffeomorphism onto its image. Let $\delta > 0$ be a Lebesgue number for the open covering $(E(V_p))_{p \in M}$ of the diagonal in the compact manifold $M \times M$, i.e., for any $q \in M$ there is a $p \in M$ with $B_\delta(q) \times B_\delta(q) \subset E(V_p)$. Hence $\operatorname{injrad}(M) \geq \delta > 0$.

$\square$

2. Almost complex and symplectic manifolds

In this section the notions of almost complex and symplectic manifolds are introduced. Almost complex manifolds are the ranges of pseudo-holomorphic curves. In this context, important examples of almost complex manifolds are given by symplectic manifolds. Basic references for this section and for further studies are [AL], [ABK], [HZ] and [MS1].

Let V denote a real vector space of finite dimension. A *complex structure* on V is an endomorphism J of V with $J^2 = -\mathbf{1}$, where $\mathbf{1}$ denotes the identity. A complex structure J on V turns V into a complex vector space by defining $(a + ib)v := av + bJv$ for $a, b \in \mathbb{R}$ and $v \in V$.

A *symplectic structure* on V is a non-degenerate, skew symmetric bilinear map $\omega \colon V \times V \to \mathbb{R}$. If ω is a symplectic structure on V, the pair (V, ω) is called a *symplectic vector space*.

Example 2.1. Let g_0 denote the standard Euclidean scalar product on $\mathbb{R}^{2m}$ and denote by $e_1, \ldots, e_{2m}$ its standard orthonormal basis. The *standard complex structure* J_0 on $\mathbb{R}^{2m}$ is defined by $J_0(e_k) = e_{m+k}$ and $J_0(e_{m+k}) = -e_k$ for $k = 1, \ldots, m$. The *standard symplectic structure* ω_0 on $\mathbb{R}^{2m}$ is given by $\omega_0(v, w) = g_0(J_0 v, w)$ for $v, w \in \mathbb{R}^{2m}$. This is equivalent to $g_0 = \omega_0 \circ (\mathbf{1} \times J_0)$.

Definition. Let V be a real vector space of finite dimension.

  (i) A complex structure J and a symplectic structure ω on V are called *compatible* if $g := \omega \circ (\mathbf{1} \times J)$ is a Euclidean scalar product on V.

  (ii) A complex structure J and a Euclidean scalar product g on V are called *compatible* if $\omega := g \circ (J \times \mathbf{1})$ is a symplectic structure on V.

 (iii) A symplectic structure ω and a Euclidean scalar product g on V are called *compatible* if the endomorphism J of V defined by $\omega = g \circ (J \times \mathbf{1})$ is a complex structure on V.

If we have compatibility under the assumption of (i), (ii) or (iii) there is an isomorphism of V to $\mathbb{R}^{2m}$, $2m = \dim V$, transforming the two given structures, and the third one which is determined by the two compatible ones, to the standard structures in $\mathbb{R}^{2m}$. To see this, note that compatibility implies that there is a g-orthonormal basis $e_1, \ldots, e_m, Je_1, \ldots, Je_m$ of (V, J, ω, g). Using this, the next proposition shows that every finite-dimensional symplectic vector space is of even dimension and symplectically isomorphic to some $\mathbb{R}^{2m}$ with its standard structure ω_0.

Proposition 2.2. *Assume (V, ω) is symplectic vector space of finite dimension. Then there exists a complex structure J on V compatible with ω.*

Proof. Choose any Euclidean scalar product g on V. Let A be the endomorphism of V defined by $g(Av, w) = \omega(v, w)$ for $v, w \in V$. Since ω is non-degenerate, A is an automorphism of V. By the skew symmetry of ω we get $A^* = -A$ for the g-adjoint map A^* of A. Hence $-A^2 = A^*A = AA^*$ is positive definite and symmetric with respect to g. Let Q be the unique positive definite square root of $-A^2$. Then $J := AQ^{-1}$ has the desired properties. In fact, since A is normal, it commutes with Q^{-1} and $J^2 = AQ^{-1}AQ^{-1} = A^2(-A^2)^{-1} = -\mathbf{1}$. Moreover, $\omega(v, Jw) = g(Av, Jw) = g(Av, AQ^{-1}w) = g(Q^{-1}A^*Av, w) = g(Qv, w)$. Since Q is symmetric and positive definite, this completes the proof.   □

Lemma 2.3. *A complex structure J and a symplectic structure ω on V are compatible if and only if the following two conditions hold:*

(i) $\omega(v, Jv) > 0$ *for each $v \in V \setminus \{0\}$.*

(ii) $\omega(Jv, Jw) = \omega(v, w)$ *for all $v, w \in V$.*

A complex structure J and a Euclidean scalar product g on V are compatible if and only if g is J-invariant, that is, $g(Jv, Jw) = g(v, w)$ for all $v, w \in V$.

Proof. This can be readily verified, using the definition.   □

Definition. A complex structure J on V is said to be *tamed* by a symplectic structure ω if (i) in the lemma holds, that is, $\omega(v, Jv) > 0$ for each $v \in V \setminus \{0\}$.

Definition. Assume J is a complex structure on V. A J-invariant Euclidean scalar product is also called a *Hermitian* scalar product on (V, J).

Let M denote a smooth manifold of finite dimension, possibly with boundary. Recall that a Riemannian metric on M is a map which assigns to each point $p \in M$ a Euclidean scalar product g_p on the tangent space T_pM in a smooth manner. Similarly, a Riemannian structure, a symplectic structure or a complex structure is defined on a real vector bundle $E \to M$ over M.

Definition. Let $E \to M$ be a smooth real vector bundle over a manifold M. A *Riemannian structure* g on the vector bundle $E \to M$ is a Euclidean scalar product g_p on each fibre E_p, $p \in M$, depending smoothly on the base point p. A *symplectic structure* ω on $E \to M$ is a symplectic structure ω_p in each E_p, depending smoothly on $p \in M$. Finally, a *complex structure* J on $E \to M$ is a complex structure J_p on each E_p, depending smoothly on p.

Depending smoothly means that, given any smooth sections X, Y of $E \to M$ the functions $g(X, Y)$, $\omega(X, Y)$ and the section JX are smooth, respectively.

The notions of *compatible, tamed* and *Hermitian* from above can be extended to corresponding structures on a vector bundle $E \to M$ by requiring that the respective statement holds in each fibre E_p.

An *almost complex structure* J and an *almost symplectic structure* ω on M is a complex structure and a symplectic structure on the tangent bundle $TM \to M$, respectively.

Thus an almost symplectic structure on M is a smooth, non-degenerate 2-form ω on M. It is called a *symplectic structure* if it is closed, i.e., its exterior derivative $d\omega$ is zero. If J is an almost complex structure on M, then the pair (M,J) is called an *almost complex manifold* and if ω is an (almost) symplectic structure, the pair (M,ω) is called (*almost*) *symplectic manifold*.

Proposition 2.2 easily generalizes to almost symplectic manifolds.

Proposition 2.4. *Assume (M,ω) is an almost symplectic manifold. Then there exists a complex structure J on M compatible with ω.*

Proof. Choose any Riemannian metric g on M and repeat the construction in the proof of Proposition 2.2 in each tangent space $T_p M$ with the scalar product g_p given by g. Note that the construction given above is unique up to the choice of g and depends smoothly on the initial data. $\qquad\square$

Remarks. 1. In the following (M,J) always denotes an almost complex manifold and (M,J,μ) an almost complex manifold with a *Hermitian metric* μ, i.e., a J-invariant Riemannian metric.

2. On each almost complex manifold (M,J) there exists a Hermitian metric. For instance, if g is a Riemannian metric on M then $g + g \circ (J \times J)$ is a Hermitian metric.

3. An almost complex structure J on M induces an orientation on M. For any complex basis $v_1,\ldots,v_m$ of any tangent space $T_p M$ define $(v_1,\ldots,v_m,Jv_1,\ldots,Jv_m)$ to be a positive oriented real basis of $T_p M$. The orientation is well-defined since any complex linear transformation of $(T_p M, J_p)$ is orientation preserving.

Example 2.5. Assume M is a complex manifold of complex dimension m and $z = (z^1,\ldots,z^m)$ is a complex-analytic chart of M. Let $z^\nu = x^\nu + iy^\nu$ denote the decomposition into real and imaginary part. If z varies in a complex-analytic atlas of M, an almost complex structure on M is well-defined by

$$J\frac{\partial}{\partial x^\nu} = \frac{\partial}{\partial y^\nu} \quad \text{and} \quad J\frac{\partial}{\partial y^\nu} = -\frac{\partial}{\partial x^\nu} \tag{2.1}$$

for $\nu = 1,\ldots,m$. This J is called the *complex structure* of the complex manifold M.

Now let (M,J) be an almost complex manifold of dimension $2m$. The almost complex structure J is called *integrable* if the differentiable structure of M can be defined by a complex-analytic atlas such that its charts z satisfy (2.1). A fundamental theorem of A. Newlander and L. Nirenberg in [NN] states that J is integrable if and only if the so-called Nijenhuis tensor $\mathcal{N}$ vanishes. It is defined by

$$\mathcal{N}(X,Y) = [X,Y] + J[JX,Y] + J[X,JY] - [JX,JY],$$

for vector fields X and Y, where $[\ ,\ ]$ denotes the Lie bracket. Since $\mathcal{N}(X,JX) = \mathcal{N}(X,X) = 0$, this implies that every 2-dimensional almost complex manifold is in fact a complex manifold and thus a Riemann surface. However, there is an elementary proof of the last fact in [MS1] and in [Sp].

Remarks. 1. If the almost complex structure of (M, J, μ) is integrable and the almost symplectic structure of M given by $\omega = \mu \circ (J \times \mathbf{1})$ is closed, and hence a symplectic structure compatible with J and μ, then (M, J, μ) is called a *Kähler manifold*. The simplest example of a Kähler manifold is any open subset of $\mathbb{C}^m \simeq \mathbb{R}^{2m}$ together with its standard structures. Other basic examples are the complex projective spaces with the Fubini-Study metric (see for instance [AL]).

2. There are lots of symplectic manifolds which do not possess any Kähler structure or any complex structure (see [Go]). In this context R. Gompf proved in [Go] that any finitely presented group can be realized as fundamental group of some 4-dimensional compact symplectic manifold.

3. Let (M, J) and (M', J') be two almost complex manifolds. There is a natural almost complex structure $J \oplus J'$ on $M \times M'$ given by $J \oplus J'(v, v') := (Jv, J'v')$ after identifying $T(M \times M')$ with $TM \times TM'$. Given Hermitian metrics μ and μ' or (almost) symplectic structures ω and ω' on (M, J) and (M', J'), respectively, corresponding structures $\mu \oplus \mu'$ or $\omega \oplus \omega'$ on $M \times M'$ are similarly defined: $\mu \oplus \mu'((v, v'), (w, w')) := \mu(v, w) + \mu'(v', w')$ and $\omega \oplus \omega'((v, v'), (w, w')) := \omega(v, w) + \omega'(v', w')$.

3. *J*-holomorphic maps

By a *surface* we always mean a 2-dimensional manifold, possibly with boundary. A surface is called *closed* if it is compact and without boundary. A surface together with an almost complex structure is a Riemann surface, since the almost complex structure is integrable.

Definition. Assume (N, j) and (M, J) are almost complex manifolds. A smooth map $f: (N, j) \to (M, J)$ is called *pseudo-holomorphic* or *j-J-holomorphic* if its differential satisfies

$$Tf \circ j = J \circ Tf. \tag{3.1}$$

This just means that $T_p f$ is complex linear for each $p \in N$. A pseudo-holomorphic map from a Riemann surface to (M, J) is also called a *J-holomorphic map*.

Every holomorphic map between complex-analytic manifolds is pseudo-holomorphic. In complex-analytic coordinates, (3.1) is equivalent to the Cauchy-Riemann equations. In general there are no non-constant solutions of (3.1) if $\dim N > 2$, not even locally (see M. Audin's article [Ad]). However, there are lots of non-trivial local solutions if (N, j) is a Riemann surface (refer to [Si]). The differential equation (3.1) is called a *generalized Cauchy-Riemann equation*.

Definition. A *pseudo-holomorphic curve* or *J-holomorphic curve* is a *J*-holomorphic map $f: S \to (M, J)$ from a Riemann surface S to an almost complex manifold (M, J). It is called *compact* and *closed* if its domain S is a compact and closed surface, respectively. A *J-holomorphic curve with boundary* is a *J*-holomorphic map $f: S \to (M, J)$, where S is a Riemann surface with non-empty boundary. A non-constant *J*-holomorphic curve $f: S^2 \to (M, J)$, where $S^2 = \mathbb{C} \cup \{\infty\}$ is the Riemann 2-sphere, is called *rational* or a *rational J-curve*.

Remarks. 1. The notion of a J-holomorphic curve generalizes the notion of a holomorphic curve in a complex manifold.

2. It can be shown that a differentiable map $D \to (M, J)$ from the open unit disc $D \subset \mathbb{C}$ satisfying the generalized Cauchy-Riemann equation is actually smooth (see, for example, [Si]).

3. In [Gr] Gromov formulates existence results for closed J-holomorphic curves in symplectic manifolds. For generic J, the space of closed J-holomorphic curves from a given Riemann surface to a symplectic manifold is in fact a manifold of finite dimension (see, for instance, [ABK], p. 117).

4. In the following chapters we study pseudo-holomorphic curves in almost complex manifolds. For our purpose the presence of a symplectic structure is not necessary. We sometimes work locally with a taming symplectic structure — it always exists locally on any almost complex manifold as (II.3.7) will show.

4. Riemann surfaces and hyperbolic geometry

In this section we put together some basic properties of Riemann surfaces, the domains of J-holomorphic maps. In this context we describe some elementary features of 2-dimensional hyperbolic geometry and sketch the proofs in such a way that the reader can complete them easily. For more details we mention [FK] for the complex analytic part, [Bu] and Chapter A of [BP] for the geometric part.

Let S be an oriented surface with Riemannian metric h. Then there exists a canonical complex structure j on S. In each tangent space it is just the rotation with angle $+\pi/2$. With respect to this complex structure, h is Hermitian. Another Riemannian metric h' on S is called *conformally equivalent* to h if there is a function $\lambda \colon S \to (0, \infty)$ such that $h = \lambda h'$. Obviously, h is conformally equivalent to h' if and only if their corresponding complex structures j and j' coincide.

Definition. Two Riemann surfaces (S, j) and (S', j') are called *conformally equivalent* and we write $(S, j) \simeq (S', j')$ if there is a biholomorphic map $\varphi \colon S \to S'$. Biholomorphic maps are also called *conformal transformations*.

Now assume $S = (S, j)$ is a connected Riemann surface without boundary. We denote by $\tilde{S} \to S$ its universal covering and by Σ the group of deck transformations. The group Σ acts properly discontinuously and freely. Recall that Σ acts *properly discontinuously* if for any compact subset $K \subset \tilde{S}$ we have $\sigma(K) \cap K \neq \emptyset$ only for finitely many $\sigma \in \Sigma$. The group Σ acts *freely* if $\sigma(z) \neq z$ for each $\sigma \in \Sigma \setminus \{\mathbf{1}\}$ and each $z \in \tilde{S}$.

The complex structure of S lifts to $\tilde{S}$. Clearly, with respect to the lifted complex structure, Σ acts by conformal transformations on $\tilde{S}$. So Σ can be viewed as a subgroup of the group

$$\operatorname{Conf}(\tilde{S}) = \left\{ \varphi \colon \tilde{S} \to \tilde{S} \mid \varphi \text{ a conformal transformation} \right\}.$$

The following fundamental theorem applies to $\tilde{S}$ for the proof of which we refer to [FK].

Uniformization Theorem. *Every simply connected Riemann surface without bound-ary is conformally equivalent either to* $\mathbb{C}$, *to the upper half-plane* $\mathbb{H} := \{ z \in \mathbb{C} \mid \mathrm{Im}\, z > 0 \}$ *or to the Riemann sphere* $S^2 = \mathbb{C} \cup \{\infty\}$.

Since each map in $\mathrm{Conf}(S^2)$ has some fixed point, $\tilde{S}$ is either $\mathbb{C}$ or $\mathbb{H}$ provided S is not conformally equivalent to S^2. The conformal transformations of $\mathbb{C}$ and $\mathbb{H}$ are

$$\mathrm{Conf}(\mathbb{C}) = \left\{ \sigma\colon \mathbb{C} \to \mathbb{C} \mid \sigma(z) = az + b,\ a, b \in \mathbb{C},\ a \neq 0 \right\},$$

$$\mathrm{Conf}(\mathbb{H}) = \left\{ \sigma\colon \mathbb{H} \to \mathbb{H} \mid \sigma(z) = \tfrac{az+b}{cz+d},\ a, b, c, d \in \mathbb{R},\ ad - bc = 1 \right\}. \qquad \textbf{(4.1)}$$

Every freely acting subgroup Σ of $\mathrm{Conf}(\mathbb{C})$ is a subgroup of the translations of $\mathbb{C}$ and so canonically a subgroup of $\mathbb{C}$ itself. If additionally Σ acts properly discontinuously on $\mathbb{C}$ it is conjugate to exactly one of the subgroups $\mathbb{Z} \oplus \tau\mathbb{Z}$, $\mathrm{Im}\,\tau > 0$, or to the sub-group $\mathbb{Z}$ of $\mathbb{C}$. Hence the connected Riemann surfaces without boundary not covered by $\mathbb{H}$ are up to conformal equivalence S^2, $\mathbb{C}$, $\mathbb{Z}\backslash\mathbb{C}$ and $\mathbb{Z} \oplus \tau\mathbb{Z}\backslash\mathbb{C}$ with $\mathrm{Im}\,\tau > 0$. These Riemann surfaces are said to be of *exceptional type*. The sphere S^2 carries a Hermit-ian metric of constant sectional curvature $+1$ and the other exceptional surfaces carry a complete flat Hermitian metric. A corresponding statement is also true for the surfaces of *non-exceptional type*, those covered by $\mathbb{H}$; they carry a complete metric of constant sectional curvature -1. In the following we shall describe the geometry behind the conformal structure of $\mathbb{H}$.

We always attach to $\mathbb{H}$ the hyperbolic metric $g_z = (\mathrm{Im}\, z)^{-2} g_{\mathrm{eucl}}$, $z \in \mathbb{H}$, where g_{eucl} denotes the standard Euclidean metric on $\mathbb{C}$. In this way $\mathbb{H}$ becomes a model of the 2-dimensional *hyperbolic plane*. It is called the *upper half-plane model*. We shall mainly deal with this model. Another useful model is the *Poincaré model*. It can be obtained by transforming $\mathbb{H}$ conformally into the open unit disc $D \subset \mathbb{C}$ via $\Phi\colon \mathbb{H} \to D$, $\Phi(z) = (z - i)(z + i)^{-1}$. The hyperbolic metric of $\mathbb{H}$ is transformed under Φ to the metric $4(1 - |z|^2)^{-2} g_{\mathrm{eucl}}$ on D.

We denote by $\mathrm{Iso}^+(\mathbb{H})$ the group of orientation preserving isometries of $\mathbb{H}$, i.e., the group of those diffeomorphisms preserving the metric and orientation of $\mathbb{H}$,

$$\mathrm{Iso}^+(\mathbb{H}) := \left\{ \sigma\colon \mathbb{H} \to \mathbb{H} \mid \sigma^* g = g,\ \sigma \text{ an orientation preserving diffeomorphism} \right\},$$

where $\sigma^* g(v, v') = g(T\sigma(v), T\sigma(v'))$.

Proposition 4.1. *The orientation preserving isometries of* $\mathbb{H}$ *are precisely its confor-mal transformations, i.e.,*

$$\mathrm{Iso}^+(\mathbb{H}) = \mathrm{Conf}(\mathbb{H}).$$

Corollary 4.2. *If S is a (possibly unconnected) Riemann surface without boundary and with all its components of non-exceptional type, then there exists a unique com-plete Hermitian metric h on S such that (S, h) and $\mathbb{H}$ are locally isometric. This metric is called the* Poincaré metric *of S.*

Remark. A Riemannian metric on S, locally isometric to $\mathbb{H}$, is called a *metric of constant curvature* -1.

Proof of the corollary. Each component of the Riemann surface S can be realized as a quotient of $\mathbb{H}$ by a properly discontinuously and freely acting subgroup of $\mathrm{Conf}(\mathbb{H}) = \mathrm{Iso}^+(\mathbb{H})$. Thus the metric of $\mathbb{H}$ passes to the quotient.   □

Proof of the proposition. First of all, by direct calculation, one readily verifies that $\sigma^* g = g$ for each $\sigma \in \mathrm{Conf}(\mathbb{H})$. Hence $\mathrm{Conf}(\mathbb{H}) \subset \mathrm{Iso}^+(\mathbb{H})$. Now the proposition is a direct consequence of the next lemma.

Lemma 4.3. $\mathrm{Conf}(\mathbb{H})$ *acts transitively on the subset of unit tangent vectors of* $T\mathbb{H}$, *i.e., given* $v, v' \in T\mathbb{H}$, $\|v\| = \|v'\| = 1$, *there is some* $\sigma \in \mathrm{Conf}(\mathbb{H})$ *with* $T\sigma(v) = v'$.

Indeed, the lemma implies that for any orientation preserving isometry $\phi \in \mathrm{Iso}^+(\mathbb{H})$, there exists some $\sigma \in \mathrm{Conf}(\mathbb{H})$ with $T_i \sigma = T_i \phi \colon T_i \mathbb{H} \to T_{\phi(i)} \mathbb{H}$. We conclude that $\phi = \sigma$ since $\psi = \exp_{\psi(i)} \circ T_i \psi \circ \exp_i^{-1}$ for any isometry ψ.   □

Proof of the lemma. The conformal transformations which fix the point $i \in \mathbb{H}$ are precisely the maps

$$z \mapsto \frac{az - b}{bz + a}, \quad a, b \in \mathbb{R}, a^2 + b^2 = 1. \tag{4.2}$$

By differentiation one sees that their differentials act transitively on the unit tangent sphere at i. Finally, one easily sees that, given any $p \in \mathbb{H}$, there exists a conformal transformation mapping i to p.   □

Definition. By $\overline{\mathbb{H}}$ we mean the closure of $\mathbb{H} \subset \mathbb{C} \cup \{\infty\}$ in the Riemann sphere. The circle $\mathbb{H}(\infty) := \overline{\mathbb{H}} \setminus \mathbb{H} = \mathbb{R} \cup \{\infty\}$ is called the *boundary* (*at infinity*) of $\mathbb{H}$ and its elements are called *points at infinity*. Note that any conformal transformation of $\mathbb{H}$ extends to a homeomorphism of $\overline{\mathbb{H}}$.

Lemma 4.4. *The unparametrized geodesics of* $\mathbb{H}$ *are the Euclidean rays parallel to the imaginary axis, i.e., the sets* $x + i\mathbb{R}_+$, $x \in \mathbb{R}$, *and the half-circles with centre on the real axis, i.e., the sets* $x + re^{i(0,\pi)}$, $x \in \mathbb{R}$, $r \in \mathbb{R}_+$, *as illustrated in Figure 1.*

Sketch of proof. The imaginary axis $i\mathbb{R}_+$, as fixed point set of the isometry $\mathbb{H} \ni z \mapsto -\bar{z}$ of $\mathbb{H}$, is an unparametrized geodesic. A direct calculation shows that the image of $i\mathbb{R}_+$ under a conformal transformation is either a ray parallel to the imaginary axis or a half-circle as in the lemma. Since $\mathrm{Conf}(\mathbb{H})$ acts transitively on the set of unit tangent vectors to $\mathbb{H}$, we see that each of the sets $x + i\mathbb{R}_+$ and $x + re^{i(0,\pi)}$ as in the lemma is the image of $i\mathbb{R}_+$ under some conformal transformation. These sets are unparametrized geodesic since $\mathrm{Iso}^+(\mathbb{H}) = \mathrm{Conf}(\mathbb{H})$. And, since each tangent vector to $\mathbb{H}$ is tangent to one of these sets, each unparametrized geodesic is of that form.   □

Remark 4.5. Where hyperbolic geometry is concerned (in this section and in Chapter IV), depending on the context, *geodesic* (*segment*) refers both to a parametrized and unparametrized geodesic (segment) and is denoted by the same letter. This should not give rise to any confusion. The closure in $\overline{\mathbb{H}}$ of any geodesic (segment) c in $\mathbb{H}$ is homeomorphic to a closed interval. Its boundary points we also call the *endpoints* of c. If any of its endpoints is contained in $\mathbb{H}(\infty)$ it is denoted by $c(-\infty)$ or $c(\infty)$, depending whether it is the left or right endpoint with respect to a chosen parametrization or at least orientation of c.

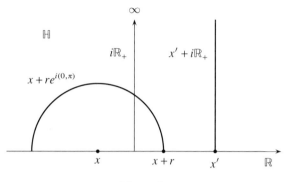

Figure 1.

The above description of geodesics shows that, given any two different points $p, q \in \mathbb{H}$, there is a unique geodesic segment connecting p to q. Hence all geodesics c in $\mathbb{H}$ are minimizing, that is $d(c(t), c(s)) = |t - s|$ if c is parametrized by arc length. Moreover, we have the following

Lemma 4.6. *For any $p \in \mathbb{H}$ the exponential map* $\exp_p \colon T_p\mathbb{H} \to \mathbb{H}$ *at p is a diffeomorphism.*

Sketch of proof. The path $\mathbb{R} \ni t \mapsto ie^t$ is a geodesic in $\mathbb{H}$ parametrized by arc length. By (4.2) the unit speed geodesics in $\mathbb{H}$ starting at i are given by

$$\mathbb{R} \ni t \mapsto \frac{aie^t - b}{bie^t + a}, \quad a, b \in \mathbb{R}, \; a^2 + b^2 = 1.$$

Applying the conformal transformation $\Phi \colon \mathbb{H} \to D$, $\Phi(z) = (z - i)(z + i)^{-1}$, one shows that in the Poincaré model the unit speed geodesics starting at 0 are $\mathbb{R} \ni t \mapsto z \tanh \frac{t}{2}$, $z \in \partial D$. Thus we see that the exponential map of the Poincaré model in 0 is a diffeomorphism. By transitivity of the isometry group of the hyperbolic plane the lemma follows. $\square$

Lemma 4.7. *The distance function d of $\mathbb{H}$ is given by*

$$\sinh\left(\tfrac{1}{2}d(z, w)\right) = \frac{|z - w|}{2\left(\mathrm{Im}(z)\,\mathrm{Im}(w)\right)^{1/2}} \qquad \text{for } z, w \in \mathbb{H}. \tag{4.3}$$

Sketch of proof. The path $\mathbb{R} \ni t \mapsto ie^t$ is a geodesic parametrized by arc length and thus $d(i, ie^t) = t$. This is consistent with formula (4.3) as claimed in the lemma. It is not difficult to show that the right side of (4.3) is $\mathrm{Iso}^+(\mathbb{H})$-invariant.

Transitivity of $\mathrm{Iso}^+(\mathbb{H})$ on the set of unit tangent vectors concludes the proof. $\qquad\square$

Lemma 4.8. *Let* $\Sigma \subset \mathrm{Iso}^+(\mathbb{H})$ *be a subgroup acting properly discontinuously and freely and denote by* Σz *the equivalence class of* $z \in \mathbb{H}$ *in* $\Sigma \backslash \mathbb{H}$. *The injectivity radius* $\mathrm{injrad}(\Sigma \backslash \mathbb{H}, \Sigma z)$ *of* $\Sigma \backslash \mathbb{H}$ *at* Σz *is given by*

$$\mathrm{injrad}(\Sigma \backslash \mathbb{H}, \Sigma z) = \tfrac{1}{2} \inf_{\sigma \in \Sigma \backslash \{1\}} d(z, \sigma(z)) .$$

This is half the length of the shortest geodesic loop in $\Sigma \backslash \mathbb{H}$ *through* Σz.

Remark. A *geodesic loop* in $\Sigma \backslash \mathbb{H}$ through Σz is a non-constant geodesic segment $c: [a, b] \to \Sigma \backslash \mathbb{H}$ with $c(a) = c(b) = \Sigma z$.

Proof. Let $\pi: \mathbb{H} \to \Sigma \backslash \mathbb{H}$ denote the canonical projection. Given $\sigma \in \Sigma \backslash \{1\}$, we consider the geodesic segment c in $\mathbb{H}$ connecting z to $\sigma(z)$. Let w be the midpoint of c, i.e., the point w in c satisfying $d(z, w) = d(w, \sigma(z)) = (1/2)d(z, \sigma(z))$. Observe that $\pi \circ c$ is a geodesic loop through Σz. Thus, there exist two different tangent vectors $v_1, v_2 \in T_{\Sigma z}(\Sigma \backslash \mathbb{H})$ with $\|v_k\| = (1/2)d(z, \sigma(z))$ and $\exp_{\Sigma z}(v_k) = \pi(w)$ for $k = 1, 2$. It follows that

$$\mathrm{injrad}(\Sigma \backslash \mathbb{H}, \Sigma z) \leq \tfrac{1}{2} \inf_{\sigma \in \Sigma \backslash \{1\}} d(z, \sigma(z)) .$$

In order to prove the converse inequality, note that $\exp_z : T_z \mathbb{H} \to \mathbb{H}$ is a diffeomorphism and that $\exp_{\Sigma z} \circ T_z \pi = \pi \circ \exp_z$. So it remains to prove that $\pi|_{B_r(z)}$ is injective for $r := (1/2) \inf_{\sigma \in \Sigma \backslash \{1\}} d(z, \sigma(z))$. Assume that there exist two different points $p, q \in B_r(z)$ with $\pi(p) = \pi(q)$. Then there exists some $\sigma \in \Sigma \backslash \{1\}$ with $\sigma(p) = q$. Consequently, $d(z, \sigma(z)) \leq d(z, q) + d(\sigma(p), \sigma(z)) = d(z, q) + d(p, z) < 2r$ in contradiction to the definition of r. $\qquad\square$

Before proceeding, we shall fix some notation.

Definition and Lemma 4.9. *Every fixed point free isometry* $\sigma \in \mathrm{Iso}^+(\mathbb{H})$ *is conjugate to exactly one of the isometries*

$$T_l : z \mapsto e^l z, \qquad l > 0 \tag{4.4}$$

or to exactly one of the isometries P, P^{-1} *with*

$$P : z \mapsto z + 1 . \tag{4.5}$$

In the first case σ *is called* hyperbolic *(with displacement* l*) and in the second case* parabolic. *We call the* T_l standard hyperbolic *isometries,* P *the* standard parabolic *isometry.*

Proof. Note that any homeomorphism of $\overline{\mathbb{H}}$ has some fixed point by Brouwer's theorem (see [Hi] or [Mi]). Hence a fixed point free isometry σ of $\mathbb{H}$ has at least one fixed point on $\mathbb{H}(\infty)$. However, if it had more than two, it would be the identity as one readily checks using (4.1). After suitable conjugation in $\mathrm{Iso}^+(\mathbb{H})$ we may assume that the fixed points of σ are precisely 0 and ∞ in case σ has two fixed points on $\mathbb{H}(\infty)$ and that the fixed point of σ is ∞ in case σ has only one fixed point. The elements in $\mathrm{Iso}^+(\mathbb{H})$ having fixed points exactly 0 and ∞ are the maps T_l, $l \in \mathbb{R} \setminus \{0\}$. Possibly interchanging the role of 0 and ∞ by conjugation with the orientation preserving isometry $z \mapsto -z^{-1}$ yields the first part of the claim. The elements in $\mathrm{Iso}^+(\mathbb{H})$ having the only fixed point ∞ are the maps $z \mapsto z + r$, $r \in \mathbb{R} \setminus \{0\}$. Conjugation with a suitable T_l, $l \in \mathbb{R}$, yields the second part of the lemma; observe that any isometry of $\mathbb{H}$ conjugating P to P^{-1} has to reverse the orientation. $\qquad\square$

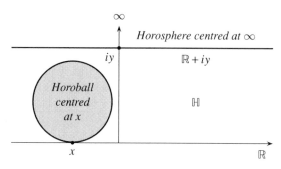

Figure 2.

Definition. The standard hyperbolic isometries T_l leave a unique geodesic, namely the imaginary axis $i\mathbb{R}_+$, invariant as a point set. From the previous lemma we obtain that each hyperbolic isometry σ leaves a geodesic invariant as a point set. It is called the *axis* of σ. The standard parabolic isometry leaves each horizontal Euclidean line $\mathbb{R} + iy$, $y > 0$, invariant; they are called *horospheres* centred at ∞. Images of horospheres centred at ∞ under an isometry σ are called horospheres centred at $\sigma(\infty)$. The horospheres centred at $x \in \mathbb{R} \subset \mathbb{H}(\infty)$ are the Euclidean circles in $\overline{\mathbb{H}}$ tangent to the line $\mathbb{R} \subset \mathbb{H}(\infty)$ in x with x removed. If HS is a horosphere in $\mathbb{H}$, the component HB of $\mathbb{H} \setminus \mathrm{HS}$ whose closure in $\overline{\mathbb{H}}$ contains exactly one point of $\mathbb{H}(\infty)$, namely the centre of HS, is called (*open*) *horoball*, the closure of HB in $\mathbb{H}$ is called *closed* horoball.

The map $\mathrm{SL}(2, \mathbb{R}) \to \mathrm{Iso}^+(\mathbb{H})$,

$$\begin{pmatrix} a & b \\ c & d \end{pmatrix} \mapsto \left\{ z \mapsto \frac{az + b}{cz + d} \right\}$$

is a surjective group homomorphism with kernel $\{\pm 1\}$ and one gets a natural isomorphism

$$\mathrm{PSL}(2, \mathbb{R}) := {\mathrm{SL}(2, \mathbb{R})}/{\{\pm 1\}} \simeq \mathrm{Iso}^+(\mathbb{H}).$$

Via this isomorphism the absolute value $|\operatorname{tr} \sigma|$ of the trace of an element $\sigma \in \mathrm{Iso}^+(\mathbb{H})$ is well-defined. Note that

$$\mathrm{PSL}(2, \mathbb{R}) \ni \begin{bmatrix} e^{l/2} & 0 \\ 0 & e^{-l/2} \end{bmatrix} \simeq T_l \quad \text{and} \quad \mathrm{PSL}(2, \mathbb{R}) \ni \begin{bmatrix} 1 & 1 \\ 0 & 1 \end{bmatrix} \simeq P. \quad \textbf{(4.6)}$$

Since the trace is invariant under conjugacy, this proves the following result.

Lemma 4.10. *Let $\sigma \in \mathrm{Iso}^+(\mathbb{H})$ be a fixed point free isometry. Then*
  (i)   *σ is hyperbolic if and only if $|\operatorname{tr}\sigma| > 2$.*
  (ii) *σ is parabolic if and only if $|\operatorname{tr}\sigma| = 2$.*
In case of (i) the displacement of σ is $2\operatorname{arccosh}((1/2)|\operatorname{tr}\sigma|) \in \mathbb{R}_+$.

Remark. Consequently, an isometry $\sigma \in \mathrm{Iso}^+(\mathbb{H}) \setminus \{1\}$ has some fixed point in $\mathbb{H}$ if and only if $|\operatorname{tr}\sigma| < 2$.

In the remarks below we describe the metric of $\mathbb{H}$ with respect to other useful coordinates on $\mathbb{H}$. The reader may wish to compute the formulas as an exercise or refer to [Kl].

Remark 4.11. We fix some $p \in \mathbb{H}$ and some unit vector $v \in T_p\mathbb{H}$. Then the map $\mathbb{R}_+ \times S^1 \to \mathbb{H} \setminus \{p\}$, where $S^1 = 2\pi\mathbb{Z}\backslash\mathbb{R}$, given by

$$(r, \theta) \mapsto \exp_p(re^{i\theta} \cdot v)$$

is a diffeomorphism. Its inverse map (r, θ): $z \mapsto (r(z), \theta(z))$ defines *polar coordinates* with origin p. With respect to these coordinates, the hyperbolic metric g of $\mathbb{H}$ is given by

$$g\left(\frac{\partial}{\partial r}, \frac{\partial}{\partial r}\right) \equiv 1, \quad g\left(\frac{\partial}{\partial \theta}, \frac{\partial}{\partial \theta}\right) = \sinh^2 r \quad \text{and} \quad g\left(\frac{\partial}{\partial r}, \frac{\partial}{\partial \theta}\right) \equiv 0.$$

Remark 4.12. Let $c \colon \mathbb{R} \to \mathbb{H}$ be a geodesic or a horosphere parametrized with constant speed $u > 0$. Then the map $\mathbb{R}^2 \to \mathbb{H}$, given by

$$(t, s) \mapsto \exp_{c(t)}\left(s\frac{i\dot{c}(t)}{u}\right)$$

is a diffeomorphism. Its inverse map (t, s): $z \mapsto (t(z), s(z))$ are so-called *Fermi coordinates* with respect to c. If c is a geodesic then

$$g\left(\frac{\partial}{\partial t}, \frac{\partial}{\partial t}\right) = u^2 \cosh^2 s, \quad g\left(\frac{\partial}{\partial s}, \frac{\partial}{\partial s}\right) \equiv 1 \quad \text{and} \quad g\left(\frac{\partial}{\partial t}, \frac{\partial}{\partial s}\right) \equiv 0.$$

And if c parametrizes a horosphere and $i\dot{c}$ points into the corresponding horoball then

$$g\left(\frac{\partial}{\partial t}, \frac{\partial}{\partial t}\right) = u^2 e^{-2s}, \quad g\left(\frac{\partial}{\partial s}, \frac{\partial}{\partial s}\right) \equiv 1 \quad \text{and} \quad g\left(\frac{\partial}{\partial t}, \frac{\partial}{\partial s}\right) \equiv 0.$$

We conclude this chapter with an application of the uniformization theorem, namely the classification of annuli up to conformal equivalence.

5. Annuli

Let A be a Riemann surface diffeomorphic to $S^1 \times (0, 1)$, $S^1 \times [0, 1)$ or $S^1 \times [0, 1]$. Then we call A an *open, semi-open* or *closed annulus*, respectively. We denote by ∂A the boundary of A, by $\partial_0 A$, $\partial_1 A$ the connected components of ∂A in case A is closed. By $\langle \sigma \rangle$ we mean the subgroup of $\mathrm{Iso}^+(\mathbb{H})$ generated by $\sigma \in \mathrm{Iso}^+(\mathbb{H})$. If σ is parabolic or hyperbolic then $\langle \sigma \rangle$ is isomorphic to $\mathbb{Z}$.

Definition and Lemma 5.1. *Every open annulus is conformally equivalent either to the* elliptic cylinder $\mathbb{Z}\backslash\mathbb{C}$, *the* parabolic cylinder $\langle P\rangle\backslash\mathbb{H}$ *or to exactly one of the* hyperbolic cylinders $\langle T_l\rangle\backslash\mathbb{H}$ *with* $l > 0$.

Proof. The fundamental group of A is isomorphic to $\mathbb{Z}$. The claim follows by the above description of $\mathrm{Conf}(\mathbb{C})$ and $\mathrm{Conf}(\mathbb{H})$ and the uniformization theorem.   □

Figure 3.

In the next two examples we give other useful descriptions of annuli.

Example 5.2. We identify S^1 with $2\pi\mathbb{Z}\backslash\mathbb{R}$ and give $S^1 \times I = 2\pi\mathbb{Z}\backslash(\mathbb{R} \times I)$ the product metric, where I is an open interval in $\mathbb{R}$. Then the following holds:

$$S^1 \times \mathbb{R} \simeq \mathbb{Z}\backslash\mathbb{C}$$
$$S^1 \times (0,\infty) \simeq \langle P\rangle\backslash\mathbb{H}$$
$$S^1 \times (0,r) \simeq \langle T_{2\pi^2/r}\rangle\backslash\mathbb{H} \quad \text{for } r > 0.$$

The first two assertions are clear, the third one can be verified as follows. We identify $\mathbb{R}^2$ with $\mathbb{C}$. Then the exponential map $z \mapsto e^z$ gives a conformal transformation from $\mathbb{R} \times (0,\pi)$ to $\mathbb{H}$. Since $S^1 \times (0,r)$ is conformally equivalent to $(2\pi^2/r)\mathbb{Z}\backslash\mathbb{R} \times (0,\pi)$ and since the $(2\pi^2/r)\mathbb{Z}$-action on $\mathbb{R} \times (0,\pi)$ transforms under the exponential map into a $\langle T_{2\pi^2/r}\rangle$-action on $\mathbb{H}$, the claim follows.

Definition. Let A be an annulus with $A \setminus \partial A$ conformally equivalent to $S^1 \times (0,r)$ for some $r \in (0,\infty]$ and $S^1 = 2\pi\mathbb{Z}\backslash\mathbb{R}$. Then $\mathrm{Mod}\,A := r$ is called the *modulus* of A.

Remark 5.3. The previous example shows that the modulus of a hyperbolic or parabolic cylinder A is

$$\mathrm{Mod}\,A = \sup\Big\{\, 2\pi^2\,\ell(c)^{-1} \;\Big|\; c \text{ is a homotopically non-trivial loop in } A \,\Big\}.$$

Example 5.4. Let D denote the open unit disc in $\mathbb{C}$. By the $2\pi i$-periodicity of the exponential map, we get a conformal transformation $\mathbb{C} \setminus \{0\} \to 2\pi i\mathbb{Z}\backslash\mathbb{C}$ and hence

$$\mathbb{C} \setminus \{0\} \simeq S^1 \times \mathbb{R}$$
$$D \setminus \{0\} \simeq S^1 \times (0,\infty)$$
$$\{z \in \mathbb{C} \mid r_1 < |z| < r_2\} \simeq S^1 \times (\log r_1, \log r_2), \quad 0 < r_1 < r_2.$$

For later purposes, we compute and estimate some moduli in the examples below.

Example 5.5. For $z \in \mathbb{H}$ we denote by $\arg z \in (0, \pi)$ the argument of z given by $z = |z| e^{i \arg z}$. Given $\varphi_1, \varphi_2 \in (0, \pi/2)$ with $\varphi_1 < \varphi_2$ we want to compute and estimate the modulus of the annulus

$$A = \langle T_l \rangle \Big\backslash \Big\{ z \in \mathbb{H} \mid \varphi_1 \leq \arg z \leq \varphi_2 \Big\} \subset \langle T_l \rangle \backslash \mathbb{H}$$

in terms of the injectivity radii $r_\nu := \text{injrad}\big(\langle T_l \rangle \backslash \mathbb{H}, \langle T_l \rangle e^{i \varphi_\nu} \big)$, $\nu = 1, 2$, of $\langle T_l \rangle \backslash \mathbb{H}$ in $\langle T_l \rangle e^{i \varphi_\nu}$. With the help of the exponential map we get, as in Example 5.2,

$$\text{Mod } A = \frac{2\pi}{l} (\varphi_2 - \varphi_1).$$

Using Lemma 4.7 and 4.8, we obtain

$$\sinh r_\nu = \sinh\Big(\tfrac{1}{2} d(e^{i \varphi_\nu}, e^l e^{i \varphi_\nu}) \Big)$$

$$= \frac{e^l - 1}{2 e^{l/2} \sin \varphi_\nu} = \frac{\sinh\big(\tfrac{1}{2} l\big)}{\sin \varphi_\nu}$$

and it follows that

$$\text{Mod } A = \frac{2\pi}{l} \left(\arcsin \frac{\sinh\big(\tfrac{1}{2} l\big)}{\sinh r_2} - \arcsin \frac{\sinh\big(\tfrac{1}{2} l\big)}{\sinh r_1} \right).$$

Consequently, we obtain the estimate

$$\text{Mod } A \geq 2\pi \frac{\sinh\big(\tfrac{1}{2} l\big)}{l} \left(\frac{1}{\sinh r_2} - \frac{1}{\sinh r_1} \right)$$

$$\geq \pi \left(\frac{1}{\sinh r_2} - \frac{1}{\sinh r_1} \right).$$

Example 5.6. Here we compute the modulus of the annulus

$$A = \langle P \rangle \Big\backslash \Big\{ z \in \mathbb{H} \mid y_1 \leq \text{Im} z \leq y_2 \Big\} \subset \langle P \rangle \backslash \mathbb{H}$$

for given $0 \leq y_1 \leq y_2$ in terms of the injectivity radii r_ν of $\langle P \rangle \backslash \mathbb{H}$ at $\langle P \rangle i y_\nu$, $\nu = 1, 2$. Since A is conformally equivalent to $S^1 \times (2\pi y_1, 2\pi y_2)$, we get from

$$\sinh r_\nu = \sinh\Big(\tfrac{1}{2} d(i y_\nu, i y_\nu + 1) \Big) = \frac{1}{2 y_\nu},$$

that the modulus of A is

$$\text{Mod } A = \pi \left(\frac{1}{\sinh r_2} - \frac{1}{\sinh r_1} \right).$$

Example 5.7. For $0 < r_1 < r_2$ we consider the annulus

$$A := \left\{ z \in \mathbb{H} \mid r_1 \leq d(i, z) \leq r_2 \right\}.$$

The conformal transformation $\Phi \colon \mathbb{H} \to D$, $\Phi(z) = (z - i)(z + i)^{-1}$, maps A to

$$\Phi(A) = \left\{ z \in D \mid \tanh \tfrac{r_1}{2} \leq |z| \leq \tanh \tfrac{r_2}{2} \right\},$$

and hence

$$\operatorname{Mod} A = \log\left(\tanh \tfrac{r_2}{2}\right) - \log\left(\tanh \tfrac{r_1}{2}\right).$$

Chapter II

Estimates for area and first derivatives

The goal of this chapter is to prove Gromov's monotonicity lemma and his Schwarz lemma for J-holomorphic curves using isoperimetric inequalities. The Gromov-Schwarz lemma is a generalization of the classical Schwarz lemma from complex analysis which states that for any holomorphic map f from the open unit disc in $\mathbb{C}$ into itself with $f(0) = 0$ its derivative at 0 is bounded from above by one. For any compact J-holomorphic curve $f: S \to (M, J)$ in a compact almost complex manifold the area of a piece of $f(S)$, cut from $f(S)$ by a small r-ball in M centred on the image $f(S)$, is estimated from below in terms of r by the monotonicity lemma.

1. Gromov's Schwarz- and monotonicity lemma

The differential $T_p f$ of a smooth map $f: N \to M$ between two Riemannian manifolds has a canonical norm $\|T_p f\| = \max \{ \|T_p f(v)\| \mid v \in T_p N, \|v\| = 1 \}$ at each point $p \in N$. We denote by $\|Tf\|$ the function $N \ni p \mapsto \|T_p f\|$. For a 1-form α on a Riemannian manifold, the function $\|\alpha\|$ is defined in the same way. As usual, (M, J, μ) denotes an almost complex manifold with Hermitian metric.

Theorem 1.1 (Gromov-Schwarz lemma). *Let $U \subset (M, J, \mu)$ be an open relatively compact subset and β a 1-form on U, such that its norm $\|\beta\|$ is bounded by a constant $C(\beta)$ from above. Assume further that its exterior derivative $d\beta$ satisfies $d\beta(v, Jv) \geq \mu(v, v)$ for each $v \in TU \subset TM$. Then the differential of every J-holomorphic map $f: \mathbb{H} \to U$ from the hyperbolic plane $\mathbb{H}$ is bounded by some constant depending only on U, J, μ and $C(\beta)$.*

Corollary 1.2 (Gromov-Schwarz lemma). *Let (M, J, μ) be compact. Then there exists an $\varepsilon_0 > 0$ and some constant $c > 0$ such that $\|Tf\| \leq c$ for every J-holomorphic map $f: \mathbb{H} \to M$ whose image $f(\mathbb{H})$ is contained in an ε_0-ball $B_{\varepsilon_0}(p)$ around some $p \in M$.*

Both versions, the theorem and the corollary, will be referred to as Gromov-Schwarz lemma.

Theorem 1.3 (Monotonicity lemma). *Let (M, J, μ) be compact. Then there are constants $\varepsilon_0, C_{\mathrm{ML}} > 0$ such that the following holds. Assume that $f: S \to (M, J, \mu)$ is a compact J-holomorphic curve with boundary, $s_0 \in S \setminus \partial S$ and $r \in (0, \varepsilon_0)$ such that $f(\partial S)$ is contained in the complement of the r-ball $B_r(f(s_0)) \subset M$. Then the area of f in $\overline{B}_r(f(s_0))$ satisfies*

$$\mathscr{A}\left(f(S) \cap \overline{B}_r(f(s_0))\right) \geq C_{\mathrm{ML}} \cdot r^2 .$$

Before proving these results, we shall define the area of maps from surfaces to Riemannian manifolds.

2. Area of *J*-holomorphic maps

Definition. If $f: S \to (M, g)$ is a smooth map from an oriented surface to a Riemannian manifold we define a 2-form $\sigma_{f^* g}$ on S by

$$\sigma_{f^* g}(v, w) := \left[g\big(Tf(v), Tf(v)\big) \, g\big(Tf(w), Tf(w)\big) - g^2\big(Tf(v), Tf(w)\big) \right]^{\frac{1}{2}}$$

for positively oriented (v, w) in any tangent space to S. Then

$$\mathscr{A}(f) := \int_S \sigma_{f^* g} \in [0, \infty]$$

is called the *area* of f. If $U \subset M$ is open or closed we also write $\mathscr{A}(f(S) \cap U)$ for $\mathscr{A}(f|_{f^{-1}(U)})$. If S is a surface with boundary ∂S, we denote by $\ell(\partial f)$ the length of the curve $\partial f := f|_{\partial S}$ and call $\ell(\partial f)$ the *length of the boundary* of f.

Remark. Suppose h is a Riemannian metric on S. Then its volume form is $\sigma_h := \sigma_{\mathrm{id}^* h}$ and the area of (S, h) is

$$\mathscr{A}(S) := \int_S \sigma_h \,.$$

Thus, if f is an immersion then $\sigma_{f^* g}$ is just the volume form of S with respect to the pull back metric $f^* g$.

Now assume that (S, h) is a Riemann surface with Hermitian metric h. Then every *J*-holomorphic map $f: (S, h) \to (M, J, \mu)$ is a *conformal map*, i.e., $f^* \mu = \alpha \cdot h$ for some function $\alpha: S \to \mathbb{R}$. Indeed, for $v \in TS$ we have $h(v, jv) = 0 = \mu(Tf(v), JTf(v)) = f^* \mu(v, jv)$ and $f^* \mu(v, v) = f^* \mu(jv, jv)$. Hence $f^* \mu = \|Tf\|^2 h$ and this yields in particular that

$$\mathscr{A}(f) = \int_S \|Tf\|^2 \, \sigma_h \,.$$

The right hand side of the previous equation is called the *energy* of f and thus, the area of a *J*-holomorphic curve is equal to its energy.

If ω is a compatible symplectic structure on (M, J, μ), in particular $\omega(v, w) = \mu(Jv, w)$, then the so-called *Wirtinger inequality* holds: for orthonormal tangent vectors $v, w \in TM$ we have $\omega(v, w) \le 1$ and equality holds if and only if $w = Jv$. Under these assumptions compact *J*-holomorphic curves in M have the following properties:

(i)   Closed *J*-holomorphic curves are absolutely area minimizing in their homology class in $H_2(M, \mathbb{R})$.

(ii) Assume that $U \subset M$ is open and $\omega|_U = d\lambda$ is exact. Then every compact *J*-holomorphic curve $f: S \to U$ with boundary is area minimizing among all smooth maps $\varphi: S \to U$ with $\varphi|_{\partial S} = f|_{\partial S}$, that is, $\mathscr{A}(f) \le \mathscr{A}(\varphi)$.

Namely, let $\varphi: S \to M$ be smooth. Then Wirtinger's inequality implies that

$$\mathscr{A}(\varphi) = \int_S \sigma_{\varphi^* \mu} \geq \int_S \varphi^* \omega$$

and equality holds if φ is J-holomorphic. Under the assumption of (ii) the right hand side is equal to $\int_{\partial S} \varphi^* \lambda$ by Stokes' theorem. The claims (i) and (ii) follow.

If M is compact and ω is a symplectic structure taming J, then there exists a constant $c \geq 1$ so that $c^{-1} \omega(v, Jv) \leq \mu(v, v) \leq c\omega(v, Jv)$ and one still gets area estimates

$$c^{-1} \int_S f^* \omega \leq \mathscr{A}(f) \leq c \int_S f^* \omega \tag{2.1}$$

for closed J-holomorphic curves f in M depending only on the homology class of f.

3. Isoperimetric inequalities for J-holomorphic maps

Definition. Let $\mathscr{F}$ be a family of maps from surfaces with boundary to a fixed Riemannian manifold (M, g). An *isoperimetric profile* for $\mathscr{F}$ is a map $\mathscr{I}: (0, \infty) \to (0, \infty)$ such that

$$\ell(\partial f) \geq \mathscr{I}(\mathscr{A}(f))$$

for any $f \in \mathscr{F}$. Such inequalities are called *isoperimetric inequalities* for $\mathscr{F}$.

The goal is now to deduce isoperimetric inequalities for certain families of J-holomorphic maps in order to prove the monotonicity lemma and the Gromov-Schwarz lemma.

Lemma 3.1. *If (M, J, μ) is compact there are constants $\varepsilon_0 > 0$ and $C > 0$ with the following property. For any compact J-holomorphic curve $f: S \to M$ with boundary, whose image $f(S)$ is contained in an r-ball $B_r(p)$ for some $r < \varepsilon_0$ and $p \in M$, we have*

$$4\pi\mathscr{A}(f) \leq (1 + Cr)\ell^2(\partial f).$$

Let us first give the idea of the proof. If r is smaller than the injectivity radius of M, the exponential map $\exp_p$ at any fixed $p \in M$ gives a diffeomorphism between the r-ball $B_r(p) \subset M$ and the Euclidean ball $B_r(0) \subset (T_pM, J_p, \mu_p)$ around 0. On the complex vector space (T_pM, J_p, μ_p) we have the standard symplectic form ω_0. Comparing $B_r(p) \subset M$ and $B_r(0) \subset T_pM$ we shall see that for sufficiently small $r > 0$ the Wirtinger inequality on $B_r(p) \subset M$ with respect to μ and $(\exp_p^{-1})^* \omega_0$ is "almost" satisfied. Thus a compact J-holomorphic curve $f: S \to B_r(p)$ with boundary is "almost" area minimizing among all differentiable maps $g: S \to B_r(p)$ with $g|_{\partial S} = f|_{\partial S}$. Area minimizing maps from a compact surface with boundary to some Euclidean space satisfy the well-known isoperimetric inequality $4\pi \cdot$ area $\leq$ (length of boundary)^2 (see Appendix A) which is "similar" to the one claimed in the lemma.

Proof of Lemma 3.1. Let $R = \text{injrad}(M, \mu)$ be the injectivity radius of M. Note that $R > 0$ by compactness of M (see Lemma I.1.1). Let $\exp_p := \exp_p|_{B_R(0)}$ denote the exponential map of M at p restricted to the R-ball $B_R(0) \subset (T_pM, J_p, \mu_p)$ where p is an arbitrary point in M fixed from now on. We identify canonically the tangent spaces to T_pM with T_pM itself. The standard symplectic form ω_0 of (T_pM, J_p, μ_p) is given by

$$\omega_0(v, w) = \mu_p(J_p v, w) \qquad \text{for } v, w \in T_pM.$$

We pull back ω_0 to $B_R(p) \subset M$ with $\exp_p^{-1}$ and get an exact symplectic form

$$\omega := (\exp_p^{-1})^* \omega_0$$

on $B_R(p)$.

By an easy compactness argument, which we postpone to the end of the proof, there is an $\varepsilon_0 \in (0, \text{injrad}(M, \mu))$, depending only on (M, J, μ), and some $C' > 0$, depending only on (M, J, μ) and ε_0, with the following property. For any $v \in T_pM$ and any $q = \exp_p(u)$, $u \in B_{\varepsilon_0}(0) \subset T_pM$ we have

$$\left| \, \|v\| - \|T_u \exp_p(v)\| \, \right| \leq C' d(p, q) \cdot \|v\| \tag{3.1}$$

$$\left| \, \|v\| - \|T_u \exp_p(v)\| \, \right| \leq C' d(p, q) \cdot \|T_u \exp_p(v)\| \tag{3.2}$$

$$\left| \, \|v\|^2 - \|T_u \exp_p(v)\|^2 \, \right| \leq C' d(p, q) \cdot \|T_u \exp_p(v)\|^2 \tag{3.3}$$

and

$$\left\| Jv - (T_u \exp_p)^{-1} J T_u \exp_p(v) \right\| \leq C' d(p, q) \cdot \|v\| . \tag{3.4}$$

After making ε_0 smaller, if necessary, we may additionally assume that

$$1 - C' \varepsilon_0 > 0 . \tag{3.5}$$

Using this, we get a constant $C'' > 0$ depending only on (M, J, μ) and ε_0 such that for any $q = \exp_p(u)$, $u \in B_{\varepsilon_0}(0) \subset T_pM$, and all orthonormal $v, w \in T_qM$

$$\omega(v, w) \leq 1 + C'' d(p, q) \tag{3.6}$$

$$1 - C'' d(p, q) \leq \omega(v, Jv) . \tag{3.7}$$

Namely, from the definition of ω we deduce that

$$\begin{aligned}
\omega(v, w) &= \omega_0 \big((T_u \exp_p)^{-1} v, (T_u \exp_p)^{-1} w \big) \\
&= \mu \big(J(T_u \exp_p)^{-1} v, (T_u \exp_p)^{-1} w \big) \\
&= \mu \big(Jv, w \big) + \mu \big(J(T_u \exp_p)^{-1} v - (T_u \exp_p)^{-1} Jv, (T_u \exp_p)^{-1} w \big) \\
&\quad + \mu \big((T_u \exp_p)^{-1} Jv, (T_u \exp_p)^{-1} w \big) - \mu \big(Jv, w \big) .
\end{aligned} \tag{3.8}$$

Using the inequalities (3.4) and (3.2), we estimate

$$\left| \mu\left(J(T_u \exp_p)^{-1}v - (T_u \exp_p)^{-1}Jv, (T_u \exp_p)^{-1}w\right) \right|$$
$$\leq C'd(p,q) \cdot \left\| (T_u \exp_p)^{-1}v \right\| \left\| (T_u \exp_p)^{-1}w \right\|$$
$$\leq C'd(p,q) \cdot (1+C')^2 d(p,q)^2 .$$

Furthermore, we deduce with (3.3) that

$$\left| \mu\left((T_u \exp_p)^{-1}Jv, (T_u \exp_p)^{-1}w\right) - \mu\left(Jv, w\right) \right|$$
$$= \tfrac{1}{2} \left| \|(T_u \exp_p)^{-1}(Jv+w)\|^2 - \|(T_u \exp_p)^{-1}Jv\|^2 - \|(T_u \exp_p)^{-1}w\|^2 \right.$$
$$\left. - \|Jv+w\|^2 + \|Jv\|^2 + \|w\|^2 \right|$$
$$\leq \tfrac{1}{2}(4C' + C' + C')d(p,q) .$$

Together with $\mu(Jv, w) \leq 1$ the inequality (3.6) is obtained from (3.8), and (3.7) follows since $\mu(Jv, Jv) = 1$.

We may assume that $\varepsilon_0 > 0$ is small enough such that $1 - C''\varepsilon_0 > 0$. Now suppose that $f: S \to B_r(p) \subset M$ is a compact J-holomorphic curve with boundary, $r \in (0, \varepsilon_0]$ and $g: S \to B_r(p)$ is a smooth map with $g|_{\partial S} = f|_{\partial S}$. Now the area of f can be compared with the area of g,

$$\mathcal{A}(g) = \int_S \sigma_{g^*\mu} \geq (1+C''r)^{-1} \int_S g^*\omega = (1+C''r)^{-1} \int_S f^*\omega$$

from (3.6) and Stokes' theorem. Since f is J-holomorphic, we can continue with (3.7):

$$\mathcal{A}(g) \geq (1+C''r)^{-1} \int_S f^*\omega \geq (1+C''r)^{-1}(1-C''r)\mathcal{A}(f) .$$

By the classical isoperimetric inequality for maps from surfaces to Euclidean space (see Appendix A) there is a g as above such that the map $\exp_p^{-1} \circ g: S \to B_r(0) \subset T_pM$ satisfies the inequality

$$4\pi\mathcal{A}(\exp_p^{-1} \circ g) \leq \ell^2(\partial(\exp_p^{-1} \circ g))$$

in (T_pM, μ_p). For such a g we can estimate its area $\mathcal{A}(g)$ from above. Inequality (3.1) yields

$$\mathcal{A}(g) = \int_S \sigma_{g^*\mu} \leq (1+C'r)^2 \int_S \sigma_{(\exp_p^{-1} \circ g)^*\mu_p} = (1+C'r)^2 \mathcal{A}(\exp_p^{-1} \circ g)$$

and hence together with (3.2) that

$$4\pi\mathcal{A}(g) \leq (1+C'r)^2 \ell^2(\partial(\exp^{-1} \circ g)) \leq (1+C'r)^4 \ell^2(\partial g) .$$

Putting everything together, we obtain that

$$4\pi\mathscr{A}(f) \leq (1 - C''r)^{-1}(1 + C''r)(1 + C'r)^4 \ell^2(\partial f).$$

since $f|_{\partial S} = g|_{\partial S}$. Thus

$$4\pi\mathscr{A}(f) \leq (1 + Cr)\ell^2(\partial f)$$

for some constant $C > 0$ depending only on (M, J, μ) since $r < \varepsilon_0$.

To conclude the proof of the lemma it remains to show (3.1)–(3.4). For $x \in M$ let $U := \bar{B}_{R/2}(x)$ and $U' := \bar{B}_{R/4}(x)$ where $R = \text{injrad}(M, \mu) > 0$. We trivialize $TU \simeq U \times \mathbb{R}^m$ with respect to an orthonormal basis field on U. Let $V := \bar{B}_{R/4}(0) \subset \mathbb{R}^m$ and $E \colon U' \times V \to U$ denote the restriction of the exponential map with respect to the chosen trivialization. Then, for $(p, u) \in U' \times V$, $v \in \mathbb{R}^m$ and $q = E(p, u)$, we see that

$$\left| \|v\| - \|D_2 E(p, u)(v)\| \right| \leq \|\mathbf{1} - D_2 E(p, u)\| \cdot \|v\| \tag{3.9}$$

$$\leq C'\|u\| \cdot \|v\| = C'd(p, q) \cdot \|v\|$$

for some constant $C' > 0$ since $D_2 E(p, 0) = \mathbf{1}$. Here D_2 denotes the differential with respect to the second factor. Now cover the compact manifold M with finitely many open $R/4$-balls in order to deduce (3.1). In the same way one shows (3.2)–(3.4). To that end note that there exists some constant $c \geq 1$ such that

$$c^{-1}\|v\| \leq \|D_2 E(p, u)(v)\| \leq c\|v\|$$

for each $(p, u) \in U' \times V$ and $v \in \mathbb{R}^m$. Then (3.2) is an immediate consequence of (3.1), and (3.3) follows together with

$$\left| \|v\|^2 - \|D_2 E(p, u)(v)\|^2 \right| = \left| \|v\| - \|D_2 E(p, u)(v)\| \right| \cdot \left(\|v\| + \|D_2 E(p, u)(v)\| \right).$$

Arguing as in (3.9) one proves (3.4). □

Remark 3.2. The estimate (3.7) in the proof of the previous lemma shows that for any compact manifold (M, J, μ) there exists an ε_0 such that the following holds. On each ε_0-ball in M there exists an exact symplectic form $\omega = d\beta$ taming J. Consequently, each closed J-holomorphic curve $f \colon S \to M$, whose image is contained in some ε_0-ball B, is constant. Indeed, let $\omega = d\beta$ be an exact symplectic form on B taming J. Since S is compact, $\mu(v, v) \leq c\omega(v, Jv)$ for some constant $c > 0$ and each v with base point in $f(S)$. Hence $\mathscr{A}(f) \leq c \int_S f^*d\beta = c \int_{\partial S} f^*\beta = 0$ since $\partial S = \varnothing$.

The proof of the monotonicity lemma is based on the above lemma.

Proof of the monotonicity lemma 1.3. For the compact manifold (M, J, μ) we choose $\varepsilon_0 \in (0, \text{injrad}(M, \mu))$ as in the previous lemma. Let $f \colon S \to M$ be a compact J-holomorphic curve with boundary satisfying the assumptions of the monotonicity lemma for some $s_0 \in S \setminus \partial S$ and some $r \in (0, \varepsilon_0)$. Namely, $f(\partial S)$ is contained in the complement of the r-ball $B_r(f(s_0)) \subset M$. Moreover, we may assume without loss of generality that S is connected.

Consider the maps $\rho\colon M \to \mathbb{R}$ given by $\rho(p) := d(p, f(s_0))$ and $\varphi := \rho \circ f$ as in the following commutative diagram:

$$
\begin{array}{ccc}
S & \xrightarrow{\ f\ } & M \\
\quad{}_{\varphi}\searrow & & \swarrow{}_{\rho = d(\,\cdot\,, f(s_0))}\quad \\
& \mathbb{R} &
\end{array}
$$

Since $\rho \circ \exp_{f(s_0)}(v) = \|v\|$ for $v \in T_{f(s_0)}M$ and $\|v\| < \varepsilon_0$, the map φ is smooth on a neighbourhood of $\varphi^{-1}(0, r]$ in S. Notice also that $[0, r]$ is contained in the image of φ since S is assumed to be connected. Let K be the subset

$$
K := \left\{ s \in \varphi^{-1}(0, r] \mid d\varphi_s = 0 \right\} \cup \varphi^{-1}(0)
$$

of S, which is indeed a closed and thus compact subset of S. Hence $\varphi(K)$ is a compact subset of the interval $[0, r]$ and has measure zero by Sard's theorem (see, for example, [Hi] or [Mi]). For $t \in [0, r]$ we define

$$
a(t) := \mathcal{A}(f|_{\varphi^{-1}[0,t]}) = \mathcal{A}\left(f(S) \cap \overline{B}_t(f(s_0))\right).
$$

If $t \in (0, r)$ is not a critical value of φ, that is $t \notin \varphi(K)$, then there exists an open neighbourhood $I_t \subset (0, r)$ of t which does not contain any critical value of φ, since $\varphi(K)$ is compact. Consequently, the sets $\varphi^{-1}(\tau)$, $\tau \in I_t$, are 1-dimensional submanifolds of $S \setminus \partial S$.

The differential of any J-holomorphic map at any given point either has rank 0 or 2. Hence there are no critical points of f in $\varphi^{-1}(I_t) \subset S$. Thus $f^*\mu$ is non-degenerate on the neighbourhood $\varphi^{-1}(I_t)$ of $\varphi^{-1}(t)$. So the gradient $\operatorname{grad}\varphi$ of φ with respect to $f^*\mu$ is well-defined on $\varphi^{-1}(I_t)$ by

$$
f^*\mu(\operatorname{grad}\varphi, \cdot) := d\varphi .
$$

Note that $\operatorname{grad}\varphi$ vanishes nowhere on I_t and $\|\operatorname{grad}\varphi\| \leq 1$. Indeed, $|f^*\mu(\operatorname{grad}\varphi, v)| = |d\varphi(v)| = |d\rho \circ Tf(v)| \leq 1$ for any v tangent to $\varphi^{-1}(I_t)$ with $f^*\mu(v, v) = \|Tf(v)\|^2 = 1$, since $\|d\rho\| = 1$.

For $\varepsilon > 0$ so small that $[t - \varepsilon, t + \varepsilon] \subset I_t$, a diffeomorphism

$$
\Phi\colon \varphi^{-1}(t) \times [t - \varepsilon, t + \varepsilon] \to \varphi^{-1}([t - \varepsilon, t + \varepsilon])
$$

is defined to be the solution of the ordinary differential equation (see, for example, [Hi] p. 153 for more details)

$$
\Phi(\cdot, t) = \operatorname{id}_{\varphi^{-1}(t)} \quad \text{and} \quad \frac{\partial \Phi}{\partial \tau}(s, \tau) = \frac{\operatorname{grad}\varphi}{\|\operatorname{grad}\varphi\|^2} \circ \Phi(s, \tau)
$$

for $(s, \tau) \in \varphi^{-1}(t) \times [t - \varepsilon, t + \varepsilon]$. Then Φ satisfies $\Phi(\varphi^{-1}(t) \times \{\tau\}) = \varphi^{-1}(\tau)$ for $\tau \in [t - \varepsilon, t + \varepsilon]$. Observe that a is differentiable at t since

$$|a(t + \delta) - a(t)| = \int_{\varphi^{-1}[t, t+\delta]} \sigma_{f^*\mu} = \int_{\varphi^{-1}(t) \times [t, t+\delta]} \Phi^* \sigma_{f^*\mu}$$

for $|\delta| < \varepsilon$. Here, by $[t, t + \delta]$ we mean the interval $[t + \delta, t]$ if $\delta < 0$. We see that

$$a'(t) = \lim_{\delta \to 0} \frac{1}{|\delta|} \int_{\varphi^{-1}(t) \times [t, t+\delta]} \Phi^* \sigma_{f^*\mu} \, .$$

And, since $\| \operatorname{grad} \varphi \| \leq 1$, this gives the estimate

$$a'(t) \geq \lim_{\delta \to 0} \frac{1}{|\delta|} \int_{\varphi^{-1}(t) \times [t, t+\delta]} (\| \operatorname{grad} \varphi \| \circ \Phi) \Phi^* \sigma_{f^*\mu}$$

$$= \lim_{\delta \to 0} \frac{1}{\delta} \int_t^{t+\delta} \ell(f|_{\varphi^{-1}(\tau)}) \, d\tau = \ell(f|_{\varphi^{-1}(t)}), \tag{3.10}$$

where $\ell(f|_{\varphi^{-1}(\tau)})$ denotes as usual the length of the curve $f|_{\varphi^{-1}(\tau)}$. By assumption, $f(\partial S)$ is contained in $M \setminus \overline{B}_t(f(s_0))$ and hence $\varphi^{-1}[0, t]$ is a submanifold of S with smooth boundary $\varphi^{-1}(t)$. Applying Lemma 3.1, we deduce from (3.10) that

$$a'(t) \geq 2\kappa \sqrt{a(t)}$$

for some positive constant κ depending only on (M, J, μ) and ε_0. Therefore the monotonically increasing function $\sqrt{a} \colon [0, r] \to \mathbb{R}$ is differentiable on the open subset $(0, r) \setminus \varphi(K) \subset [0, r]$ and satisfies

$$\sqrt{a}' \geq \kappa > 0. \tag{3.11}$$

Since $\varphi(K)$ is a closed set of measure zero, this implies that

$$\sqrt{a(r)} \geq \kappa r,$$

and hence the claim. Namely, let $I_v = (b_v, c_v)$, $v = 1, 2, 3, \ldots$, be the countable possibly finite number of connected components of $(0, r) \setminus \varphi(K)$. Since $\sqrt{a}$ is monotonically increasing, we deduce from (3.11) that

$$\sqrt{a(r)} \geq \sum_v \left(\sqrt{a(c_v)} - \sqrt{a(b_v)} \right) \geq \kappa \sum_v (c_v - b_v) = \kappa r,$$

and this finishes the proof of the monotonicity lemma. $\square$

Remark 3.3. Using the fact (see [Fe] 2.9.19) that any monotonically non-decreasing function $h \colon [a, b] \to \mathbb{R}$ is almost everywhere differentiable with $h' \in L^1([a, b], \mathbb{R})$ satisfying $h(b) - h(a) \geq \int_a^b h'(t) \, dt$, and applying the coarea formula ([Fe] 3.2.11), one could argue more directly in the last proof.

With the help of the monotonicity lemma, the statement of Lemma 3.1 can even be sharpened.

Lemma 3.4. *For the compact manifold (M, J, μ) choose ε_0 such that the assumptions of Lemma 3.1 and of the monotonicity lemma are satisfied. Let $f: S \to M$ be a compact J-holomorphic curve with boundary and connected domain S. Assume that its diameter $\delta(f) := \sup\{ d(f(s), f(s')) \mid s, s' \in S \}$ is smaller than ε_0. Then f satisfies the isoperimetric inequality*

$$4\pi \mathscr{A}(f) \leq (1 + c_n \ell(\partial f)) \ell^2(\partial f)$$

for some constant $c_n > 0$ depending only on (M, J, μ), ε_0 and the number n of boundary components of S.

Proof. Let $f: S \to M$ be a compact J-holomorphic curve with boundary and connected domain S. Suppose that $\delta(f) < \varepsilon_0$. Then for any $s \in S$ we have

$$d(f(s), f(\partial S)) \leq \sqrt{\frac{\mathscr{A}(f)}{C_{\mathrm{ML}}}}.$$

Otherwise there would exist an $s \in S$ such that $f(\partial S)$ would not intersect some r-ball $B_r(f(s))$ in M with $d(f(s), f(\partial S)) > r > \sqrt{\mathscr{A}(f)/C_{\mathrm{ML}}}$. This would contradict the monotonicity lemma. Applying Lemma 3.1 we get

$$d(f(s), f(\partial S)) \leq c' \ell(\partial f)$$

for any $s \in S$ and some constant c' depending only on (M, J, μ) and ε_0.

We denote by $\partial_1 S, \ldots, \partial_n S$ the connected components of ∂S. Then we have for any $s, s' \in S$ that

$$d(f(s), f(s')) \leq 2n(c' + 1) \ell(\partial f) =: c'' \ell(\partial f).$$

Indeed, choose some point $s_\nu \in \partial_\nu S$ in each component of ∂S. Obviously the balls $B_{(c'+1)\ell(\partial f)}(f(s_\nu))$, $\nu = 1, \ldots, n$, cover $f(S)$ and therefore $f(S) \subset B_{c'' \ell(\partial f)}(f(s))$ for any $s \in S$ since S is connected. Together with Lemma 3.1 we get

$$4\pi \mathscr{A}(f) \leq \min\left\{ (1 + C\varepsilon_0) \ell^2(\partial f), (1 + Cc'' \ell(\partial f)) \ell^2(\partial f) \right\}.$$

$\square$

The previous lemma stated an isoperimetric inequality for compact J-holomorphic curves with boundary in compact manifolds provided their images were contained in sufficiently small balls. Next, it is shown that the same isoperimetric inequality holds provided only the area of the compact J-holomorphic curve with boundary is sufficiently small.

Lemma 3.5. *There is an $\varepsilon_n > 0$ depending only on (M, J, μ), ε_0 and the number n of components of ∂S such that the following holds. For any compact J-holomorphic curve $f: S \to M$ with boundary and connected domain S satisfying $\mathscr{A}(f) < \varepsilon_n$, the isoperimetric inequality from the previous lemma holds with the same constant c_n.*

Proof. Assume that $f: S \to M$ is a compact J-holomorphic curve with boundary and connected domain S. Suppose that the area of f is smaller than $C_{\mathrm{ML}} \varepsilon_0^2$. After applying the monotonicity lemma as in the last proof it follows that

$$d(f(s), f(\partial S)) \le \sqrt{\frac{\mathcal{A}(f)}{C_{\mathrm{ML}}}} < \varepsilon_0 \qquad \text{for any } s \in S. \tag{3.12}$$

Arguing indirectly, we assume that

$$4\pi \mathcal{A}(f) > (1 + c_n \ell(\partial f)) \ell^2(\partial f). \tag{3.13}$$

In particular, this implies that $\ell(\partial f) < \sqrt{4\pi \mathcal{A}(f)}$. Using the same covering argument as at the end of the previous proof together with (3.12), the following estimate for the diameter of f is obtained,

$$\delta(f) \le 2n \left(\sqrt{C_{\mathrm{ML}}^{-1} \mathcal{A}(f)} + \sqrt{4\pi \mathcal{A}(f)} \right) =: r.$$

We see from (3.13) that f does not satisfy the assumption of the previous lemma and consequently, $r \ge \varepsilon_0$. This implies that

$$\mathcal{A}(f) \ge \varepsilon_n > 0 \quad \text{with} \quad \varepsilon_n = \frac{\varepsilon_0^2}{4n^2} \left(\sqrt{C_{\mathrm{ML}}^{-1}} + \sqrt{4\pi} \right)^{-2}.$$

This ε_n has the desired properties. □

The next lemma gives area estimates for compact J-holomorphic curves with boundary whose image is contained in an open subset U satisfying the assumptions of the Gromov-Schwarz lemma 1.1.

Lemma 3.6. *Let $U \subset M$ be an open subset of a manifold (M, J, μ), β a 1-form on U such that its norm is bounded by a positive constant $C(\beta)$ and whose exterior derivative satisfies $d\beta(v, Jv) \ge \mu(v, v)$ for any $v \in TU \subset TM$. Then $\mathcal{A}(f) \le C(\beta) \cdot \ell(\partial f)$ for any compact J-holomorphic curve $f: S \to U \subset M$ with boundary.*

Proof. This lemma is an immediate consequence of Stokes' theorem,

$$\mathcal{A}(f) = \int_S \sigma_{f^* \mu} \le \int_S f^* d\beta = \int_{\partial S} f^* \beta \le C(\beta) \cdot \ell(\partial f).$$

□

4. Proof of the Gromov-Schwarz lemma

Let $U \subset (M, J, \mu)$ and β be as in the assumption of Theorem 1.1, i.e., U is open and relatively compact and β is a 1-form on U. The norm of β is bounded on U by some constant $C(\beta)$ and $d\beta$ satisfies $d\beta(v, Jv) \ge \mu(v, v)$ for each $v \in TU$. Using the results from the last section, we shall see later that a J-holomorphic map $f: D \to U$ satisfies the assumption of the lemma below. It gives an estimate for the norm of the differential at zero of conformal maps f from the open unit disc $D \subset \mathbb{C}$ to a Riemannian manifold provided the maps $f|_{\overline{D}_r}$, $r \in (0,1)$ and $\overline{D}_r := \{z \in \mathbb{C} \mid |z| \leq r\}$, satisfy a certain isoperimetric inequality. Then the Gromov-Schwarz lemma 1.1 follows immediately.

Lemma 4.1. *Let (M, g) be a Riemannian manifold and $\mathcal{I}\colon (0,\infty) \to (0,\infty)$ a continuous map satisfying*

$$(i) \quad \int_0^1 \left(\frac{1}{\mathcal{I}^2(t)} - \frac{1}{4\pi t} \right) dt < \infty \qquad and \qquad (ii) \quad \int_1^\infty \frac{dt}{\mathcal{I}^2(t)} < \infty .$$

Assume that $f\colon D \to (M, g)$ is a conformal map with $\ell(\partial(f|_{\overline{D}_r})) \geq \mathcal{I}(\mathcal{A}(f|_{\overline{D}_r}))$ for any $r \in (0,1)$. Then $\|T_0 f\| \leq c(\mathcal{I})$ for some constant $c(\mathcal{I})$ depending only on $\mathcal{I}$.

Proof. Let f be as in the lemma and assume that $T_0 f \neq 0$. We define

$$a(r) := \mathcal{A}(f|_{\overline{D}_r}), \quad l(r) := \ell(\partial(f|_{\overline{D}_r})) \quad \text{and} \quad \varepsilon(t) := \frac{1}{\mathcal{I}^2(t)} - \frac{1}{4\pi t}.$$

With respect to usual polar coordinates (ρ, θ) on D we have that

$$a(r) = \iint_{\overline{D}_r} \|Tf\|^2 \, \rho \, d\rho \, d\theta$$

since f is conformal. Under the assumption of the lemma, the derivative of a can be estimated using the Cauchy-Schwarz inequality:

$$a'(r) = \int_0^{2\pi} \|Tf\|^2 \, r \, d\theta \geq \frac{1}{2\pi r} \left(\int_0^{2\pi} \|Tf\| \, r \, d\theta \right)^2 = \frac{1}{2\pi r} l^2(r) \geq \frac{1}{2\pi r} \mathcal{I}^2(a(r)) .$$

Dividing by $\mathcal{I}^2(a(r))$ and integrating this inequality, we obtain that

$$\phi(a(r_2)) - \phi(a(r_1)) \geq \frac{1}{2\pi} \log \frac{r_2}{r_1} \qquad \text{for } 0 < r_1 < r_2 < 1,$$

where ϕ is a primitive of $\mathcal{I}^{-2}$, say

$$\phi(s) = \int_1^s \frac{dt}{\mathcal{I}^2(t)} = \frac{1}{4\pi} \log s + \int_1^s \varepsilon(t) \, dt .$$

We now estimate

$$\int_1^{a(r_2)} \frac{dt}{\mathcal{I}^2(t)} = \int_1^{a(r_1)} \frac{dt}{\mathcal{I}^2(t)} + \int_{a(r_1)}^{a(r_2)} \frac{dt}{\mathcal{I}^2(t)}$$

$$\geq \frac{1}{4\pi} \log a(r_1) + \int_1^{a(r_1)} \varepsilon(t) \, dt + \frac{1}{2\pi} \log \frac{r_2}{r_1} .$$

Rearranging the right hand side yields that

$$\int_1^{a(r_2)} \frac{dt}{\mathcal{I}^2(t)} \geq \frac{1}{4\pi} \log \frac{a(r_1)}{\pi r_1^2} + \frac{1}{2\pi} \log r_2 + \frac{1}{4\pi} \log \pi + \int_1^{a(r_1)} \varepsilon(t) \, dt . \qquad (4.1)$$

Now observe that $a(r)/\pi r^2$ converges to $\|T_0 f\|^2$ as $r \to 0$ since

$$\left| a(r) - \|T_0 f\|^2 \pi r^2 \right| = \left| \iint_{D_r} \left(\|Tf\|^2 - \|T_0 f\|^2 \right) \rho \, d\rho \, d\theta \right|$$

$$\leq \sup_{D_r} \left| \|Tf\|^2 - \|T_0 f\|^2 \right| \cdot \iint_{D_r} \rho \, d\rho \, d\theta$$

$$= \sup_{D_r} \left| \|Tf\|^2 - \|T_0 f\|^2 \right| \cdot \pi r^2 .$$

Taking the limit $r_2 \to 1$ and $r_1 \to 0$ in (4.1), properties (i), (ii) of $\mathscr{I}$ imply the result.

$\square$

The next remark is important for later purposes.

Remark 4.2. The statement of Lemma 4.1 also holds under a weakened assumption on f. Namely, instead of assuming f to be conformal, it suffices to require only that f is a continuous map $D \to (M, g)$, differentiable and conformal on $D \setminus \{z_0\}$ for some point $z_0 \neq 0$. Furthermore, the area of f in some neighbourhood of z_0 should be finite. Under these assumptions on f the proof still holds. We only have to be a little bit careful in one step: with the map a,

$$a(r) = \iint_{\overline{D}_r \setminus \{z_0\}} \|Tf\|^2 \rho \, d\rho \, d\theta ,$$

$\phi \circ a$ is also continuous on $[0, 1)$, differentiable on $(0, 1) \setminus \{|z_0|\}$ and monotonically increasing. Then the integral

$$\int_{r_1}^{r_2} \frac{a'(r)}{\mathscr{I}^2(a(r))} \, dr = \int_{r_1}^{r_2} (\phi \circ a)'(r) \, dr$$

exists and has the value $\phi(a(r_2)) - \phi(a(r_1))$.

Now the proof of the Gromov-Schwarz lemma can be concluded. For that purpose we shall put together the isoperimetric inequalities from Lemma 3.5 and 3.6 to get an isoperimetric profile $\mathscr{I}: (0, \infty) \to (0, \infty)$ for J-holomorphic maps from a closed disc to U which has properties (i) and (ii) in Lemma 4.1. Observe that U is relatively compact in (M, J, μ) and thus the isoperimetric inequalities from above hold in U, with constants depending on U and (M, J, μ).

Let $\psi: (0, \infty) \to (0, \infty)$ be the strictly monotonically increasing map given by $\psi(l) = (4\pi)^{-1} (1 + c_1 l) l^2$. Then define

$$\mathscr{I}(t) := \begin{cases} \psi^{-1}(t), & \text{if } t < \frac{\varepsilon_1}{2} \\ \chi(t), & \text{if } \frac{\varepsilon_1}{2} \leq t \leq \varepsilon_1 \\ C(\beta)^{-1} \cdot t, & \text{if } t > \varepsilon_1 , \end{cases}$$

where χ is chosen such that $\chi(t) \leq \max\{C(\beta)^{-1} \cdot t, \psi^{-1}(t)\}$ and $\mathscr{I}$ becomes continuous. Here c_1 and ε_1 are constants as in Lemma 3.5 for compact J-holomorphic curves in U with one boundary component. Hence $\mathscr{I}$ is an isoperimetric profile for J-holomorphic curves from a closed disc to U and it depends only on U, (M,J,μ) and $C(\beta)$; χ can be given explicitly depending only on ε_1, c_1 and $C(\beta)$. It is evident that $\mathscr{I}$ satisfies the assumption (ii) of Lemma 4.1,

$$\int_1^\infty \frac{dt}{\mathscr{I}^2(t)} < \infty.$$

We show that it also satisfies (i):

$$4\pi t = 4\pi \psi \circ \psi^{-1}(t) = (1 + c_1 \mathscr{I}(t))\mathscr{I}^2(t) \qquad \text{for } 0 < t \le \frac{\varepsilon_1}{2},$$

and thus

$$\frac{1}{\mathscr{I}^2(t)} - \frac{1}{4\pi t} = \frac{c_1 \mathscr{I}(t)}{4\pi t} = \frac{c_1 \psi^{-1}(t)}{4\pi t} \qquad \text{for } 0 < t \le \frac{\varepsilon_1}{2}.$$

Substituting $t = \psi(l)$, we see that

$$\int_0^{\varepsilon_1/2} \left(\frac{1}{\mathscr{I}^2(t)} - \frac{1}{4\pi t} \right) dt < \infty.$$

Now apply Lemma 4.1 to get an upper bound for $\|T_0 f\|$ depending only on U, J, μ and $C(\beta)$. Moreover, with respect to the Poincaré metric λ on D, there is an orientation preserving isometry mapping any given $z \in D$ to 0. Thus, the same estimate holds for the differential at any point $z \in D$ with respect to λ as it holds at 0. Since the open unit disc D with the Poincaré metric is isometric to $\mathbb{H}$, this completes the proof of the Gromov-Schwarz lemma 1.1.

Proof of the Gromov-Schwarz lemma 1.2. The goal is to show that for any compact manifold (M,J,μ) there exists an $\varepsilon_0 > 0$ such that the differential of any J-holomorphic map $f: \mathbb{H} \to M$ whose image is contained in some ε_0-ball $B_{\varepsilon_0}(p) \subset M$ is uniformly bounded.

We identify $(T_p M, J_p, \mu_p) = \mathbb{R}^{2m}$ with its standard structures. Then the exterior derivative of $\alpha_0 := \sum_{\nu=1}^m x^\nu dy^\nu$ is $\omega_0 = \sum_{\nu=1}^m dx^\nu \wedge dy^\nu$ on $(T_p M, J_p, \mu_p)$. On $B_{\varepsilon_0}(0) \subset T_p M$ the 1-form α_0 is bounded by a constant depending only on ε_0. From the proof of Lemma 3.1, especially (3.2) and (3.7), we see that $(\exp^{-1})^* \alpha_0 =: \alpha$ is bounded on $B_{\varepsilon_0}(p) \subset M$ by some constant depending only on (M,J,μ) and ε_0. Moreover, there is a constant κ, also only depending on (M,J,μ) and ε_0, such that $\beta := \kappa\alpha$ satisfies the assumption of the Gromov-Schwarz lemma 1.1 on $U = B_{\varepsilon_0}(p) \subset M$. $\square$

Remark 4.3. As a final remark we point out that for any compact manifold (M,J,μ) an $\varepsilon_0 > 0$ can be chosen such that the following is true.

(i) (M,J,μ) together with ε_0 satisfies the assumption of the Gromov-Schwarz lemma 1.2 and of the monotonicity lemma 1.3.

(ii) There is no non-constant closed J-holomorphic curve in M whose image is contained in some ε_0-ball in M.

Indeed, (i) is clear and (ii) follows from Remark 3.2.

Chapter III

Higher order derivatives

This chapter is concerned with the estimate of higher order derivatives of J-holomorphic maps. For that purpose the differentials of J-holomorphic maps are made into pseudo-holomorphic maps to which the Gromov-Schwarz lemma applies. This is used in the proof of Gromov's theorem on the removal of singularities for J-holomorphic maps. It is a generalization of a theorem of Riemann from complex analysis, which says that a holomorphic map $f: S \setminus \{a\} \to S^2$ from a Riemann surface minus an interior point a to the Riemann sphere can be extended to a holomorphic map $S \to S^2$, provided f does not have an essential singularity at a. As another application it can be proved that the derivatives of a locally uniformly convergent sequence of pseudo-holomorphic maps also converge locally uniformly. This generalizes a theorem of Weierstraß for holomorphic functions.

1. 1-jets of J-holomorphic maps

The object of this section is to turn the differentials of J-holomorphic maps, i.e., their 1-jets, into pseudo-holomorphic maps. The reader may also refer to P. Gauduchon's approach in [AL]. The approach here is similar to the one suggested by M.-P. Muller in [Mu].

Let (S, j) be a Riemann surface and (M, J) an almost complex manifold of dimension $2m$. We define

$$[S, M] := \bigcup_{(s,p) \in S \times M} \mathrm{Hom}_{\mathbb{C}}(T_s S, T_p M),$$

to be the disjoint union of the complex vector spaces of complex linear homomorphisms $T_s S \to T_p M$. This is canonically the total space of a complex vector bundle

$$[S, M] \xrightarrow{\pi} S \times M$$

over $S \times M$ with fibre $[S, M]_{(s,p)} = \mathrm{Hom}_{\mathbb{C}}(T_s S, T_p M)$ over $(s, p) \in S \times M$. A local trivialization is given at the beginning of the proof of Lemma 1.2.

In this section let $f: G \to M$ always denote a J-holomorphic map from an open subset $G \subset S$ to M.

Definition. The map $f^{(1)}: G \to [S, M]$ given by $s \mapsto T_s f \in [S, M]_{(s, f(s))}$ for $s \in G$ is called the 1-*jet* of f. It is a lift of the graph $\Gamma_f: G \to S \times M$ of f to $[S, M]$, that means $\pi \circ f^{(1)} = \Gamma_f$.

Our aim is to show that there is an almost complex structure on the manifold $[S,M]$ such that the 1-jet $f^{(1)}$ of any J-holomorphic map $f: G \to M$, $G \subset S$ open, is again pseudo-holomorphic.

We can view every point $\tau \in [S,M]_{(s,p)}$ as a complex one-dimensional linear subspace $\widehat{\tau} := \{ v + \tau(v) \mid v \in T_s S \}$ of $T_s S \oplus T_p M = T_{(s,p)}(S \times M)$. By

$$\mathcal{E}_\tau := T\pi^{-1}(\widehat{\tau})$$

we denote the preimage of $\widehat{\tau}$ under $T\pi: T[S,M] \to T(S \times M)$ and define

$$\mathcal{E} := \bigcup_{\tau \in [S,M]} \mathcal{E}_\tau.$$

Lemma 1.1. *The image of $Tf^{(1)}: TG \to T[S,M]$ is in fact contained in $\mathcal{E}$.*

Proof. $T\pi \circ Tf^{(1)}(T_s S) = TT_f(T_s S) = \widehat{f^{(1)}(s)}$. $\square$

Let $\mathcal{V}[S,M] \to [S,M]$ denote the *vertical bundle* of $[S,M] \xrightarrow{\pi} S \times M$, that is,

$$\mathcal{V}[S,M] = T\pi^{-1}(0)$$

is the preimage of the zero section of $T(S \times M)$ under $T[S,M] \xrightarrow{T\pi} T(S \times M)$. Actually $\mathcal{V}[S,M]$ is a complex vector bundle, canonically isomorphic to the vector bundle $\pi^*[S,M] \to [S,M]$ obtained by pulling back the bundle $[S,M] \xrightarrow{\pi} S \times M$ with π itself, namely

$$\pi^*[S,M] = \left\{ (\tau,\sigma) \in [S,M] \times [S,M] \mid \pi(\tau) = \pi(\sigma) \right\},$$

and the projection to the base space is the projection onto the first factor. Then the canonical isomorphism is given by

$$\pi^*[S,M] \ni (\tau,\sigma) \mapsto \frac{d}{dt}\Big|_{t=0}(\tau + t\sigma) \in \mathcal{V}[S,M]. \tag{1.1}$$

Note that we have

$$\pi^*[S,M] \simeq \mathcal{V}[S,M] \subset \mathcal{E} \subset T[S,M].$$

Moreover,

Lemma 1.2. *The subset $\mathcal{E} \subset T[S,M]$ is in fact a linear subbundle.*

Proof. Let $z = x + iy$ be a holomorphic chart on an open subset $G \subset S$ and $\xi = (x^1, \dots, x^{2m})$ a chart on an open subset $V \subset M$. Denote by $\psi: TV \to \mathbb{R}^{2m}$ the fibrewise linear isomorphism given by

$$\psi\left(\sum_{v=1}^{2m} a^v \frac{\partial}{\partial x^v}\Big|_p \right) = (a^1, \dots, a^{2m}).$$

Consequently, the map $\chi \colon \pi^{-1}(G \times V) \to \mathbb{R}^{2m}$, defined by

$$\chi(\tau) = \psi \circ \tau\left(\frac{\partial}{\partial x}\bigg|_s\right)$$

with $\pi(\tau) = (s, p) \in G \times V$, gives a trivialization of $[S, M] \to S \times M$ over $G \times V$ and hence

$$\eta \colon \pi^{-1}(G \times V) \to z(G) \times \xi(V) \times \mathbb{R}^{2m}$$

$$\eta(\tau) = \big((z, \xi) \circ \pi(\tau), \chi(\tau)\big)$$

is a chart for the manifold $[S, M]$ defined on $\pi^{-1}(G \times V)$. We denote the components of η by

$$\eta = (\tilde{x}, \tilde{y}, \tilde{x}^1, \ldots, \tilde{x}^{2m}, r^1, \ldots, r^{2m}).$$

Then X and Y, given by

$$X(\tau) = \frac{\partial}{\partial \tilde{x}}\bigg|_\tau + \sum_{\alpha=1}^{2m} r^\alpha(\tau) \frac{\partial}{\partial \tilde{x}^\alpha}\bigg|_\tau$$

$$Y(\tau) = \frac{\partial}{\partial \tilde{y}}\bigg|_\tau + \sum_{\alpha,\beta=1}^{2m} J_\beta^\alpha(p) r^\beta(\tau) \frac{\partial}{\partial \tilde{x}^\alpha}\bigg|_\tau,$$

are local sections of $T[S, M] \to [S, M]$ defined on $\pi^{-1}(G \times V)$ where $\pi(\tau) = (s, p)$ and the J_β^α are the components of the almost complex structure J of M with respect to the local basis field $\partial/\partial x^1, \ldots, \partial/\partial x^{2m}$. Observe that $X(\tau), Y(\tau) \in \mathscr{E}_\tau$ since

$$T\pi X(\tau) = \frac{\partial}{\partial x}\bigg|_s + \sum_{\alpha=1}^{2m} r^\alpha(\tau) \frac{\partial}{\partial x^\alpha}\bigg|_p = \frac{\partial}{\partial x}\bigg|_s + \tau\left(\frac{\partial}{\partial x}\bigg|_s\right) \in \hat{\tau} \tag{1.2}$$

and similarly

$$T\pi Y(\tau) = \frac{\partial}{\partial y}\bigg|_s + J \circ \tau\left(\frac{\partial}{\partial x}\bigg|_s\right) = \frac{\partial}{\partial y}\bigg|_s + \tau\left(\frac{\partial}{\partial y}\bigg|_s\right) \in \hat{\tau}. \tag{1.3}$$

Notice also that $X(\tau), Y(\tau) \notin \mathscr{V}[S, M]_\tau$ since $T\pi(X(\tau))$ and $T\pi(Y(\tau))$ are both non-zero. Moreover, $X(\tau)$ and $Y(\tau)$ are linearly independent since their images under $T\pi$ are linearly independent. Actually $\partial/\partial r^1, \ldots, \partial/\partial r^{2m}$ give a local real basis field for the vertical bundle $\mathscr{V}[S, M] \to [S, M]$ and thus $X, Y, \partial/\partial r^1, \ldots, \partial/\partial r^{2m}$ is a family of point-wise linearly independent smooth vector fields on $\pi^{-1}(G \times V) \subset [S, M]$ spanning $\mathscr{E}_\tau$ at $\tau \in \pi^{-1}(G \times V)$ since $\dim_\mathbb{R} \mathscr{E}_\tau = 2 + \dim_\mathbb{R} \mathscr{V}[S, M]_\tau$. This proves the lemma. $\qquad\square$

We denote the complex structure of the vector bundle $\mathscr{V}[S, M] \to [S, M]$ by $\tilde{J}$. Note that, with respect to the basis field $\partial/\partial r^1, \ldots, \partial/\partial r^{2m}$ as in the previous proof, the complex structure $\tilde{J}$ of the vertical bundle is given by

$$\tilde{J}_\beta^\alpha(\tau) = J_\beta^\alpha(p), \quad \pi(\tau) = (s, p).$$

This complex structure can be extended to $\mathscr{E}$, as the next proposition shows.

Proposition 1.3. *There exists a complex structure $J^{(1)}$ on $\mathscr{E} \to [S, M]$ having the following properties:*

(i) $J^{(1)}$ *is an extension of $\tilde{J}$, that is $J^{(1)}|_{\mathscr{V}[S,M]} = \tilde{J}$.*

(ii) *For each open subset $G \subset S$ and each J-holomorphic map $f \colon G \to M$, the differential $Tf^{(1)} \colon TG \to \mathscr{E}$ is fibrewise complex linear.*

With respect to this $J^{(1)}$, the map $T\pi|_{\mathscr{E}} \colon \mathscr{E} \to T(S \times M)$ is fibrewise complex linear.

For the sake of completeness we make the following

Addition to Proposition 1.3. *Properties (i) and (ii) determine $J^{(1)}$ uniquely.*

The proof of the addition below is not self-contained, but we shall not make use of the uniqueness result.

Proof of uniqueness. Assuming $J^{(1)}$ satisfying (i) and (ii) exists, then it is unique, provided there exist sufficiently many locally defined J-holomorphic maps. And in fact there are. Namely, for any $\tau \in [S, M]$ there is a J-holomorphic map f defined in some neighbourhood of $s \in S$, where $\pi(\tau) = (s, p)$, with $f^{(1)}(s) = \tau$. For this local existence result we refer to Theorem 3.1.1 in [Si]. For such an f condition (ii) means that

$$J^{(1)}Tf^{(1)}(v) = Tf^{(1)}(jv) \quad \text{for } v \in T_s S. \tag{1.4}$$

Together with (i), that is

$$J^{(1)}|_{\mathscr{V}[S,M]} = \tilde{J}, \tag{1.5}$$

this determines $J^{(1)}$, if it exists, uniquely. $\square$

Proof of Proposition 1.3. We pick $\tau \in [S, M]$ and assume that such a J-holomorphic map f with $f^{(1)}(s) = \tau$ exists. Then we have to define $J_\tau^{(1)}$ by (1.4) and (1.5) and to show that this definition is actually independent of the choice of f. Adapting the notion from above, we work in local coordinates (z, ξ) of $S \times M$ near $\pi(\tau) = (s, p)$. The generalized Cauchy-Riemann equations are given in such coordinates by

$$f_y^\alpha = \sum_{\beta=1}^{2m} (J_\beta^\alpha \circ f) f_x^\beta, \tag{1.6}$$

where f_x^α and f_y^α denote the derivatives of the components $f^\alpha = x^\alpha \circ f$ in direction $\partial/\partial x$ and $\partial/\partial y$, respectively. Differentiation of (1.6) with respect to x yields

$$f_{yx}^\alpha = \sum_{\beta,v=1}^{2m} \left((J_{\beta,v}^\alpha \circ f) f_x^v f_x^\beta + (J_\beta^\alpha \circ f) f_{xx}^\beta \right). \tag{1.7}$$

The components of $\eta \circ f^{(1)}$ are

$$\eta \circ f^{(1)} = (x, y, f^1, \ldots, f^{2m}, f_x^1, \ldots, f_x^{2m})$$

and with respect to the chosen coordinates, the differential of $f^{(1)}$ can be described,

$$Tf^{(1)}\frac{\partial}{\partial x} = \frac{\partial}{\partial \tilde{x}} + \sum_{\alpha=1}^{2m} f_x^\alpha \frac{\partial}{\partial \tilde{x}^\alpha} + \sum_{\alpha=1}^{2m} f_{xx}^\alpha \frac{\partial}{\partial r^\alpha}$$

$$Tf^{(1)}\frac{\partial}{\partial y} = \frac{\partial}{\partial \tilde{y}} + \sum_{\alpha=1}^{2m} f_y^\alpha \frac{\partial}{\partial \tilde{x}^\alpha} + \sum_{\alpha=1}^{2m} f_{xy}^\alpha \frac{\partial}{\partial r^\alpha}.$$

Inserting (1.6) and (1.7) into the last equation yields that

$$Tf^{(1)}\frac{\partial}{\partial y} = \frac{\partial}{\partial \tilde{y}} + \sum_{\alpha,\beta=1}^{2m} (J_\beta^\alpha \circ f) f_x^\beta \frac{\partial}{\partial \tilde{x}^\alpha}$$
$$+ \sum_{\alpha,\beta,\nu=1}^{2m} \left((J_{\beta,\nu}^\alpha \circ f) f_x^\nu f_x^\beta + (J_\beta^\alpha \circ f) f_{xx}^\beta \right) \frac{\partial}{\partial r^\alpha}.$$

We are now able to compute $J_\tau^{(1)} X(\tau)$, where

$$X = \frac{\partial}{\partial \tilde{x}} + \sum_{\alpha=1}^{2m} r^\alpha \frac{\partial}{\partial \tilde{x}^\alpha}$$

is as above, namely

$$J_\tau^{(1)} X(\tau) = J_\tau^{(1)} \left(Tf^{(1)} \left(\frac{\partial}{\partial x}\Big|_s \right) - \sum_{\alpha=1}^{2m} f_{xx}^\alpha(p) \frac{\partial}{\partial r^\alpha}\Big|_\tau \right)$$

$$= Tf^{(1)} \left(\frac{\partial}{\partial y}\Big|_s \right) - \sum_{\alpha,\beta=1}^{2m} J_\beta^\alpha(p) f_{xx}^\beta(p) \frac{\partial}{\partial r^\alpha}\Big|_\tau$$

$$= \frac{\partial}{\partial \tilde{y}}\Big|_\tau + \sum_{\alpha,\beta=1}^{2m} J_\beta^\alpha(p) f_x^\beta(p) \frac{\partial}{\partial \tilde{x}^\alpha}\Big|_\tau$$
$$+ \sum_{\alpha,\beta,\nu=1}^{2m} J_{\beta,\nu}^\alpha(p) f_x^\nu(p) f_x^\beta(p) \frac{\partial}{\partial r^\alpha}\Big|_\tau$$

$$= \frac{\partial}{\partial \tilde{y}}\Big|_\tau + \sum_{\alpha,\beta=1}^{2m} J_\beta^\alpha(p) r^\beta(\tau) \frac{\partial}{\partial \tilde{x}^\alpha}\Big|_\tau$$
$$+ \sum_{\alpha,\beta,\nu=1}^{2m} J_{\beta,\nu}^\alpha(p) r^\nu(\tau) r^\beta(\tau) \frac{\partial}{\partial r^\alpha}\Big|_\tau \qquad (1.8)$$

and this is independent of the choice of f with $f^{(1)}(s) = \tau$. Thus we can define a complex structure $J^{(1)}$ on the vector bundle $\mathscr{E} \to [S, M]$ by (1.8) and (1.5) which is indeed smooth. It follows also from equation (1.8) that

$$T\pi J^{(1)} X(\tau) = \frac{\partial}{\partial y}\Big|_s + \tau \left(\frac{\partial}{\partial y}\Big|_s \right)$$

and consequently we obtain that $T\pi|_{\mathscr{E}}$ is fibrewise complex linear, indeed

$$T\pi J^{(1)}X(\tau) = (j \oplus J)\left(\frac{\partial}{\partial x}\Big|_s + \tau\left(\frac{\partial}{\partial x}\Big|_s\right)\right)$$
$$= (j \oplus J)T\pi X(\tau).$$

This finishes the proof of the proposition. $\square$

The next goal is to extend the complex structure $J^{(1)}$ of $\mathscr{E} \to [S,M]$ to a complex structure of the tangent bundle $T[S,M] \to [S,M]$ so that $[S,M]$ becomes an almost complex manifold. The extension will not be unique.

Choose a linear subbundle $\mathscr{H}$ of $T[S,M]$ such that $T[S,M]$ splits into the direct sum

$$T[S,M] = \mathscr{V}[S,M] \oplus \mathscr{H}.$$

For instance, let $\mathscr{H}_\tau$ be the orthogonal complement of $\mathscr{V}[S,M]_\tau$ in $T_\tau[S,M]$ with respect to some Riemannian metric on $[S,M]$. Then $\mathscr{H} \cap \mathscr{E}$ is a real 2-dimensional linear subbundle of $T[S,M]$. Namely, decompose the vector fields X and $J^{(1)}X$ from the last proof into their components with respect to $\mathscr{V}[S,M] \oplus \mathscr{H}$, that is $X = X_\mathscr{V} + X_\mathscr{H}$ and $J^{(1)}X = (J^{(1)}X)_\mathscr{V} + (J^{(1)}X)_\mathscr{H}$. It is evident that $X_\mathscr{H}$ and $(J^{(1)}X)_\mathscr{H}$ are linearly independent and define a local trivialization of $\mathscr{H} \cap \mathscr{E}$.

Pulling back a Hermitian metric h on $(S \times M, j \oplus J)$ with $T\pi|_\mathscr{H}: \mathscr{H} \to T(S \times M)$ gives a Riemannian structure on the vector bundle $\mathscr{H} \to [S,M]$ since $T\pi|_\mathscr{H}$ is fibrewise an isomorphism. Now build the orthogonal complement $\mathscr{K}$ of $\mathscr{H} \cap \mathscr{E}$ in $\mathscr{H}$ with respect to this Riemannian structure. Then, a complex structure on $\mathscr{K}$ is well-defined by pulling back the complex structure $j \oplus J$ of $T(S \times M) \to S \times M$ with $T\pi|_\mathscr{K}$. Indeed, h is Hermitian and $(j \oplus J)(T\pi X_\mathscr{K}) = T\pi(J^{(1)}X_\mathscr{K})$ which implies that $T\pi(\mathscr{K}_\tau)$ is a complex linear subspace of $T_{\pi(\tau)}(S \times M)$.

In this manner the complex structure $J^{(1)}$ of $\mathscr{E} \to [S,M]$ is extended to a complex structure on $T[S,M] = \mathscr{E} \oplus \mathscr{K} \to [S,M]$. We also denote such a complex structure by $J^{(1)}$. This shows the following

Proposition 1.4. *There exists an almost complex structure $J^{(1)}$ on $[S,M]$ such that for any pseudo-holomorphic map $f: G \to M, G \subset S$ open, the 1-jet $f^{(1)}: G \to ([S,M], J^{(1)})$ is again pseudo-holomorphic.*

Later the Gromov-Schwarz lemma will be applied to 1-jets in order to estimate derivatives of second order. Passing successively to 1-jets, the same can be done for higher order derivatives.

Definition. Let (S, j) be a Riemann surface and (M, J) an almost complex manifold. For $n = 0, 1, 2, \dots$ we define inductively an almost complex manifold $([S,M]^{(n)}, J^{(n)})$ by $([S,M]^{(0)}, J^{(0)}) := (M, J)$ and

$$\left([S,M]^{(n+1)}, J^{(n+1)}\right) := \left([S, [S,M]^{(n)}], (J^{(n)})^{(1)}\right).$$

We construct $J^{(n+1)}$ from $j \oplus J^{(n)}$ in the same way we constructed $J^{(1)}$ from $j \oplus J = j \oplus J^{(0)}$. The $J^{(n)}$ are not uniquely determined. Hence, if (S, j) and (M, J) are given, we choose them once and leave them fixed. For any open subset $G \subset S$ and any J-holomorphic map $f: G \to M$ we define $f^{(0)} := f: G \to [S, M]^{(0)}$ and inductively pseudo-holomorphic maps

$$f^{(n+1)} := (f^{(n)})^{(1)}: G \to \left([S, M]^{(n+1)}, J^{(n+1)}\right).$$

We call $f^{(n)}$ the n-jet of the J-holomorphic map f.

2. Removal of singularities

In this section Gromov's theorem on the removal of singularities is stated and proved.

Theorem 2.1 (Removal of singularities). *Let S be a Riemann surface, (M, J) an almost complex manifold, $a \in S$ an interior point and $f: S \setminus \{a\} \to (M, J)$ a J-holomorphic map with relatively compact image in M. Assume that f has one of the following properties with respect to some Hermitian metric μ on M.*

(i)   *There exists a neighbourhood G of a in S such that $f|_{G \setminus \{a\}}$ has finite area.*

(ii) *There exists a neighbourhood G of a in S and an open neighbourhood U of $f(G \setminus \{a\})$ in M such that the following holds. There exists a bounded 1-form β on U satisfying $d\beta(v, Jv) \geq \mu(v, v)$ for each $v \in TU \subset TM$.*

Then f can be extended to a J-holomorphic map on S.

Observe that the statement is purely local. So it is sufficient to prove the theorem for $S = G = D$ the open unit disc in $\mathbb{C}$ and $a = 0$. Let $f: D \setminus \{0\} \to (M, J, \mu)$ be a J-holomorphic map with relatively compact image in M satisfying at least one of the assumptions (i), (ii) of Theorem 2.1 with $S = D = G$ and $a = 0$. The proof has essentially two steps, namely:

(a) There is a continuous extension of f to D.

(b) Assume that f satisfies (a). Then its 1-jet $f^{(1)}: D \setminus \{0\} \to ([D, M], J^{(1)})$ satisfies the assumption of the theorem on the removal of singularities and thus has a continuous extension to D.

From (b) we get immediately that f is differentiable at 0 with $df_0 = f^{(1)}(0)$. Especially df_0 is complex linear. Applying (b) successively implies that the n-jets $f^{(n)}$, $n > 1$, are also continuously extensible to 0 and hence the extension $f: D \to M$ is smooth.

Proof of (a). We have to consider two cases according to the assumptions (i) and (ii).

CASE 1. The area of f is finite. As usual we denote by D_r the subset $\{z \in \mathbb{C} \mid |z| < r\}$ of $\mathbb{C}$. Then there exists a sequence $(z_n)_{n \geq 1}$ in $D \setminus \{0\}$ with

$$\lim_{n \to \infty} z_n = 0 \quad \text{and} \quad \lim_{n \to \infty} \ell(\partial(f|_{\overline{D}_{|z_n|}})) = 0,$$

otherwise there would exist an $\varepsilon > 0$ and an $r_0 \in (0,1)$ such that $\ell(\partial(f|_{\overline{D}_r})) \geq \varepsilon$ for each $r \leq r_0$ and this would yield that

$$\mathscr{A}(f|_{D_{r_0} \setminus \{0\}}) = \iint_{D_{r_0} \setminus \{0\}} \|Tf\|^2 \, \rho \, d\rho \, d\theta$$

$$\geq \int_0^{r_0} (2\pi\rho)^{-1} \left(\int_0^{2\pi} \|Tf\| \, \rho \, d\theta \right)^2 d\rho$$

$$= \frac{1}{2\pi} \int_0^{r_0} \frac{\ell^2(\partial(f|_{\overline{D}_\rho}))}{\rho} \, d\rho \geq \frac{\varepsilon^2}{2\pi} \int_0^{r_0} \frac{d\rho}{\rho} = \infty \, .$$

Since $f(D \setminus \{0\})$ is relatively compact in M, we may suppose that $(f(z_n))$ converges in M, say

$$\lim_{n \to \infty} f(z_n) =: p \in M \, .$$

Assume there is another sequence $(w_n)_{n \geq 1}$ in $D \setminus \{0\}$ with $\lim w_n = 0$ and $\lim f(w_n) = q \neq p$. For $r_n := |z_n|$ and $s_n := |w_n|$ we may assume without loss of generality that

$$r_1 > s_1 > r_2 > s_2 > \cdots > r_n > s_n > r_{n+1} > s_{n+1} \cdots \, .$$

Let $\rho_0 := d(p,q)$ be the distance of p and q in M. Then choose $n_0 \geq 1$ such that for each $n \geq n_0$

$$d(f(z_n), p) < \frac{\rho_0}{8} \, , \tag{2.1}$$

$$d(f(w_n), q) < \frac{\rho_0}{8} \, , \tag{2.2}$$

$$\ell(\partial(f|_{\overline{D}_{r_n}})) < \frac{\rho_0}{8} \tag{2.3}$$

and thus for each $n \geq n_0$

$$B_{\rho_0/2}(f(w_n)) \cap f(\partial \overline{D}_{r_v}) = \varnothing \qquad \text{for } v = n, n+1 \, .$$

Since $f(D \setminus \{0\})$ is relatively compact in M, the monotonicity lemma implies that there exists some positive constant $c > 0$ such that

$$\mathscr{A}(f|_{\overline{D}_{r_n} \setminus D_{r_{n+1}}}) \geq c$$

for each $n \geq n_0$. Hence $\mathscr{A}(f) = \infty$ in contradiction to our assumption. Thus f can be extended to a continuous map $D \to M$ by defining $f(0) := p$.

CASE 2. There is a neighbourhood U of $f(D \setminus \{0\})$ in M and a 1-form β with bounded norm on U satisfying $d\beta(v, Jv) \geq \mu(v,v)$ for each $v \in TU \subset TM$. Let λ be the Poincaré metric on $D \setminus \{0\}$. We identify $(D \setminus \{0\}, \lambda)$ with $\langle P \rangle \backslash \mathbb{H}$ where P is the standard parabolic isometry $z \mapsto z + 1$ of the hyperbolic space $\mathbb{H}$. Denote by $\pi \colon \mathbb{H} \to \langle P \rangle \backslash \mathbb{H}$ the canonical projection. The Gromov-Schwarz lemma yields that $\|T(f \circ \pi)\|$ is bounded on $\mathbb{H}$. Hence $\|Tf\|$, with respect to the Poincaré metric on $D \setminus \{0\}$, is also bounded since π is locally an isometry. Since the area of $\langle P \rangle \setminus \{ z \in \mathbb{H} \mid \text{Im}(z) \geq r \}$ is finite for $r > 0$, this implies that f has finite area on some neighbourhood of 0 and we are in the situation of Case 1. This finishes the proof of (a). $\qquad\square$

Observe that the following result was obtained above.

Lemma 2.2. *A J-holomorphic map $f: S \setminus \{a\} \to (M, J)$ with relatively compact image satisfying assumption (ii) of the theorem on the removal of singularities also satisfies assumption (i) and thus has finite area in some neighbourhood of a.*

Before proving (b), we extend the statements of Lemma II.3.5 and II.3.6.

Lemma 2.3. *Let $K \subset M$ be a compact subset of (M, J, μ). Then there are constants $\varepsilon, c > 0$ with the following property. If S is a connected, compact Riemann surface with exactly one boundary component, $a \in S \setminus \partial S$ and $f: S \setminus \{a\} \to (M, J, \mu)$ is a J-holomorphic map with $f(S \setminus \{a\}) \subset K$ and $\mathcal{A}(f) < \varepsilon$, then*

$$4\pi \mathcal{A}(f) \leq (1 + c\, \ell(\partial f)) \ell^2(\partial f).$$

Lemma 2.4. *Let $U \subset (M, J, \mu)$ be open and relatively compact in M. Assume there is a 1-form β on U whose norm is bounded by some constant $C(\beta)$ from above. Moreover, we suppose that $d\beta(v, Jv) \geq \mu(v, v)$ for each $v \in TU \subset TM$. Then, for any compact Riemann surface S with boundary, any $a \in S \setminus \partial S$ and any J-holomorphic map $f: S \setminus \{a\} \to U$, we have*

$$\mathcal{A}(f) \leq C(\beta) \cdot \ell(\partial f).$$

Proof of 2.3 and 2.4. In both cases f has finite area. From the last proof we see that there is a sequence Δ_n of closed neighbourhoods of a in S, where each Δ_n is diffeomorphic to a closed disc, such that

$$\bigcap_{n \geq 1} \Delta_n = \{a\} \quad \text{and} \quad \lim_{n \to \infty} \ell(\partial(f|_{\Delta_n})) = 0.$$

This implies that

$$\mathcal{A}(f) = \lim_{n \to \infty} \mathcal{A}(f|_{S \setminus \text{int}\,\Delta_n}) \quad \text{and} \quad \ell(\partial f) = \lim_{n \to \infty} \ell(\partial(f|_{S \setminus \text{int}\,\Delta_n})),$$

where $\text{int}\,\Delta_n$ denotes the interior of Δ_n. Applying Lemma II.3.5 and II.3.6, respectively, to the maps $f|_{\Delta \setminus \text{int}\,\Delta_n}$ and taking the limit $n \to \infty$, the lemmas follow. $\qquad\square$

Proof of (b). Recall that the J-holomorphic map $f: D \setminus \{0\} \to (M, J, \mu)$ is supposed to have relatively compact image in M and to satisfy one of the assumptions of the theorem on the removal of singularities with $S = G = D$ and $a = 0$. After extending f to a continuous map on D, we may assume that f satisfies both assumptions (i) and (ii) of the theorem on the removal of singularities. Namely, suppose that f satisfies (i). Then choose a sufficiently small neighbourhood U of $f(0)$ in M such that there exists a 1-form β with bounded norm on U satisfying $d\beta(v, Jv) \geq \mu(v, v)$ for each $v \in TU \subset TM$

(see the proof of the Gromov-Schwarz lemma II.1.2 at the end of Chapter II). By continuity of f there is a neighbourhood of 0 in D whose image under f is contained in U. Without loss of generality we may assume that this neighbourhood is in fact D. If f satisfies (ii) it follows from Lemma 2.2 that f has finite area in some neighbourhood of 0.

For $w \in \mathbb{C}$ we denote by $D_{1/2}(w)$ the open disc in $\mathbb{C}$ with centre w and radius $1/2$. Let w be any point in $D_{1/2} \setminus \{0\}$ and consider the map $f|_{D_{1/2}(w)}$. Using Lemma 2.3 and 2.4 and Remark II.4.2, we can argue as in the proof of the Gromov-Schwarz lemma in Section II.4 in order to deduce the following result. The differential of $f|_{D_{1/2}(w)}$ at w is bounded by some constant C, independent of the choice of $w \in D_{1/2} \setminus \{0\}$. In other words, the differential Tf of f is bounded on $D_{1/2} \setminus \{0\}$ by C.

This shows that there exists a compact subset $K \subset [D,M]_{(0,f(0))}$ such that for any neighbourhood V of K in $[D,M]$ there exists a neighbourhood G' of 0 in D with $f^{(1)}(G' \setminus \{0\}) \subset V$. We choose a Hermitian metric $\mu^{(1)}$ on the almost complex manifold $([D,M], J^{(1)})$ and view $[D,M]_{(0,f(0))}$ as a submanifold of $[D,M]$. Additionally we require that the induced metric on $[D,M]_{(0,f(0))}$ comes from a Hermitian scalar product on the vector space $[D,M]_{(0,f(0))}$. Denote by $\mathcal{N}$ the normal bundle of $[D,M]_{(0,f(0))}$ and by $\exp^{\mathcal{N}}$ the exponential map of $([D,M], J^{(1)}, \mu^{(1)})$ restricted to some neighbourhood in $\mathcal{N}$ of the zero section of $\mathcal{N}$. Together with $T([D,M]_{(0,f(0))})$, the vector bundle $\mathcal{N}$ is also a complex subbundle of $T[D,M]|_{[D,M]_{(0,f(0))}}$ since $\mu^{(1)}$ is Hermitian. Hence we can choose an orthonormal complex basis field $X_1, \ldots, X_{m+1}$ of sections of $\mathcal{N} \to [D,M]_{(0,f(0))}$ since $[D,M]_{(0,f(0))}$ is contractible. Now identify

$$\left([D,M]_{(0,f(0))}, J^{(1)}|_{[D,M]_{(0,f(0))}}, \mu^{(1)}|_{[D,M]_{(0,f(0))}}\right) = \mathbb{C}^m$$

with its standard structures and denote by $z^v = x^v + iy^v$ the standard coordinates of $\mathbb{C}^m$. Then $z^v, X_k, J^{(1)}X_k$ with $v = 1, \ldots, m$, $k = 1, \ldots, m+1$ define Fermi coordinates $\zeta^1, \ldots, \zeta^{2m}, \eta^1, \ldots, \eta^{2m+2}$ on some neighbourhood W of K in $[D,M]$ in the following way. For $\tau \in [D,M]_{(0,f(0))}$ define

$$\zeta^v \left(\exp^{\mathcal{N}} \sum_{l=1}^{m+1} \left(t_l X_l(\tau) + s_l J^{(1)} X_l(\tau)\right)\right) = x^v(\tau)$$

$$\zeta^{v+m} \left(\exp^{\mathcal{N}} \sum_{l=1}^{m+1} \left(t_l X_l(\tau) + s_l J^{(1)} X_l(\tau)\right)\right) = y^v(\tau)$$

$$\eta^k \left(\exp^{\mathcal{N}} \sum_{l=1}^{m+1} \left(t_l X_l(\tau) + s_l J^{(1)} X_l(\tau)\right)\right) = t_k$$

$$\eta^{k+m+1} \left(\exp^{\mathcal{N}} \sum_{l=1}^{m+1} \left(t_l X_l(\tau) + s_l J^{(1)} X_l(\tau)\right)\right) = s_k$$

for $v = 1, \ldots, m$ and $k = 1, \ldots, m+1$. Indeed, by differentiation one sees that these are local coordinates near each point in K. Since $\exp^{\mathcal{N}}$, restricted to the zero section, is injective and K is compact we get coordinates on some neighbourhood W of K. With respect to these coordinates, we have for any $\tau \in [D, M]_{(0, f(0))} \cap W$ that

$$\frac{\partial}{\partial \eta^k}\bigg|_{\tau} = X_k(\tau) \quad \text{and} \quad \frac{\partial}{\partial \eta^{k+m+1}}\bigg|_{\tau} = J^{(1)} X_k(\tau).$$

Let α be the 1-form

$$\alpha := \sum_{v=1}^{m} \zeta^v d\zeta^{v+m} + \sum_{k=1}^{m+1} \eta^k d\eta^{k+m+1}.$$

Then obviously

$$d\alpha(v, J^{(1)}v) = \mu^{(1)}(v, v)$$

for any $v \in T[D, M]|_{[D,M]_{(0, f(0))}}$. Since K is compact, there exists a positive constant $c \in (0, 1]$ and a sufficiently small neighbourhood V of K in $[D, M]$ such that

$$d\alpha(v, J^{(1)}v) \geq c\mu^{(1)}(v, v)$$

for each $v \in TV \subset T[D, M]$. Moreover, α is bounded by some constant $C(\alpha)$ on V. Hence the map $f^{(1)}: D \setminus \{0\} \to ([D, M], J^{(1)})$ together with the 1-form $\beta = c^{-1}\alpha$ satisfies assumption (ii) of Theorem 2.1 on the removal of singularities. This concludes the proof of (b) and thus the proof of Theorem 2.1. $\qquad \square$

3. Converging sequences of J-holomorphic maps

Using the arguments from the proof of (b) in the last section, a theorem of Weierstraß from complex analysis can be generalized.

Proposition 3.1 (generalized Weierstraß theorem). *Let S be a Riemann surface without boundary and (M, J) an almost complex manifold. Assume $f_n: S \to M, n \geq 1$, is a sequence of J-holomorphic maps converging in the C^0-topology to a map $f: S \to M$. Then $(f_n)_{n \geq 1}$ converges even in the C^∞-topology and its limit f is in fact J-holomorphic.*

Remark. By the C^k-*topology* for some integer $k \geq 0$ we always mean the topology of locally uniform C^k-convergence. Convergence in the C^∞-*topology* means convergence in all C^k-topologies. The C^k-topology is briefly described in Appendix B.

Proof. Let $(f_n)_{n \geq 1}$ and f be as in the assumptions of the proposition. It is enough to show that the corresponding 1-jets $(f_n^{(1)})_{n \geq 1}$ also converge in C^0. Then, by induction, $(f_n^{(k)})_{n \geq 1}$ converges for any integer $k > 1$ in C^0. This implies the convergence of $(f_n)_{n \geq 1}$ in the C^∞-topology and also that f is J-holomorphic.

Choose a Hermitian metric h on S and one on M. By the C^0-convergence of (f_n), for any $s \in S$ and any neighbourhood $U \subset M$ of $f(s)$ in M there is a neighbourhood G of s in S such that $f_n(G) \subset U$ for each sufficiently large n. With Gromov's Schwarz lemma $\|T_{s'} f_n\|$ can be estimated on some smaller neighbourhood G' of s for each sufficiently large n, independent of n and the choice $s' \in G'$ from above. Indeed, we may assume that G is conformally equivalent to an open disc. With respect to the Poincaré metric λ of G, the differential Tf_n is bounded on G independent of n by the Gromov-Schwarz lemma. If G' is relatively compact in G then there is some constant $c > 1$ such that $c^{-1}h(v, v) \le \lambda(v, v) \le ch(v, v)$ for any tangent vector v with base point in G'.

Now choose a Hermitian metric $\mu^{(1)}$ on $[S, M]$. Arguing as in the final part of the last section, we see that there is a compact subset K of the fibre $[S, M]_{(s, f(s))}$ and some open neighbourhood V of K in $[S, M]$ with the following properties:

(i)   There is a bounded 1-form β on V with $d\beta(v, J^{(1)}v) \ge \mu^{(1)}(v, v)$ for each $v \in TV \subset T[S, M]$.

(ii) For a sufficiently small neighbourhood $G'' \subset G'$ of s in S we have $f_n^{(1)}(G'') \subset V$ for each sufficiently large n.

Again by Gromov's Schwarz lemma, $\|T_{s'} f_n^{(1)}\|$ is bounded for each sufficiently large n on some smaller neighbourhood $G''' \subset G''$, independent of n and the choice of $s' \in G'''$. This implies the claim (see Proposition B.3 in Appendix B).   □

Corollary 3.2. *Let $G \subset \mathbb{R}^2$ be an open ball. Assume $(\mu_n)_{n \ge 1}$ is a sequence of Riemannian metrics on G such that each (G, μ_n) is isometric to an open subset of the 2-dimensional hyperbolic space $\mathbb{H}$. Suppose that (μ_n) converges in the C^∞-topology to a Riemannian metric μ. Let j_n, j denote the complex structures on G induced by μ_n, μ, respectively. Then any C^0-convergent sequence $f_n: (G, j_n) \to (M, J)$, $n \ge 1$, of j_n-J-holomorphic maps converges in fact in C^∞ to a j-J-holomorphic map.*

Proof. Under the assumption of the corollary (G, μ) is also isometric to an open subset of the hyperbolic space. The statement to be proved is purely local. Fix an arbitrary point $s \in G$ and choose $r > 0$ such that the exponential map $\exp_s$ of (G, μ) at s is defined on the $3r$-ball around 0 in (T_sG, μ_s). Let (I_n) be a sequence of orientation preserving orthogonal transformations $I_n: (T_sG, \mu_s) \to (T_sG, \mu_{n,s})$ converging to the identity and denote by $\exp_s^n$ the exponential map of (G, μ_n) at s.

Then, for each sufficiently large n the map $\varphi_n = \exp_s^n \circ I_n \circ (\exp_s)^{-1}|_{B_{2r}(s)}$ is a well-defined isometry $B_{2r}(s) \to B_{2r}^n(s) \subset G$ of the $2r$-ball in (G, μ) around s onto the $2r$-ball $B_{2r}^n(s)$ in (G, μ_n). Thus it is also a j-j_n-biholomorphic map. Since (μ_n) converges in C^∞, the sequence (φ_n) converges in C^∞ to the inclusion $B_{2r}(s) \hookrightarrow G$ and by Lemma B.2 in Appendix B, $(\varphi_n^{-1}|_{B_r(s)})$ converges also in C^∞ to the inclusion $B_r(s) \hookrightarrow G$. The proposition implies that $(f_n \circ \varphi_n)$ is C^∞-convergent to a j-J-holomorphic map f. Hence $(f_n|_{B_r(s)}) = (f_n \circ \varphi_n \circ \varphi_n^{-1}|_{B_r(s)})$ converges in C^∞ to $f|_{B_r(s)}$ by Lemma B.1.   □

4. Variable almost complex structures

More generally, one may wish to consider sequences of pseudo-holomorphic maps in a manifold M with respect to any almost complex structure taken from a C^∞-convergent sequence of almost complex structures on M. In fact, from our point of view there is no essential difference to the case when the almost complex structure on M is fixed, which is shown below. Thus the reader may wish to skip this section.

Assume (M, J, μ) is compact and $(J_n)_{n \geq 1}$ is a sequence of almost complex structures on M converging in C^∞ to J. Then $\mu_n := \frac{1}{2}(\mu + \mu \circ (J_n \times J_n))$ is a Hermitian metric on (M, J_n) converging in C^∞ to μ as $n \to \infty$. Let $\varepsilon_0(n)$, $c(n)$ and $C_{\mathrm{ML}}(n)$ be the corresponding constants from the Gromov-Schwarz lemma II.1.2 and monotonicity lemma II.1.3. For each sufficiently large n, and thus for all n, we may assume that $\varepsilon_0(n) = \varepsilon_0$, $c(n) = c$ and $C_{\mathrm{ML}}(n) = C_{\mathrm{ML}}$, independently of n. This can easily be seen from the proofs in the last chapter. If we are interested in getting uniform estimates for higher derivatives of J_n-holomorphic maps one difficulty arises. Namely, the space $[S, M]$ of 1-jets depends on J_n as a point set.

Assume now that $S =: G \subset \mathbb{R}^2$ is an open subset of $\mathbb{R}^2$ and that (j_n) is a sequence of C^∞-converging complex structures on G with limit j. We denote by $\mathrm{pr}^*TM \to G \times M$ the pull back of the tangent bundle $TM \to M$ with respect to the projection $\mathrm{pr}: G \times M \to M$ onto the second factor. The vector bundle $\mathrm{pr}^*TM \to G \times M$ can be identified with $G \times TM \to G \times M$, where the fibre over $(s, p) \in G \times M$ is $\{s\} \times T_pM$. We pick a nowhere vanishing vector field Z on $\mathbb{R}^2$, say $Z := \partial/\partial x$. This defines the vector bundle isomorphisms

$$\chi_n: [(G, j_n), (M, J_n)] \to \mathrm{pr}^*TM, \quad \chi_n(\tau) = (s, \tau(Z_p))$$
$$\chi: [(G, j), (M, J)] \to \mathrm{pr}^*TM, \quad \chi(\tau) = (s, \tau(Z_p))$$

where (s, p) is the base point of τ.

In fact, when the almost complex structure $J^{(1)}$ was constructed on the space $[G, M] = [(G, j), (M, J)]$ we worked in local coordinates defined by such a bundle isomorphism χ composed with a trivialization of pr^*TM. With respect to χ_n and χ identify $[(G, j_n), (M, J_n)]$ with pr^*TM and $[(G, j), (M, J)]$ with pr^*TM, respectively. Let $\mathscr{E}_n$ and $\mathscr{E}$ denote the corresponding bundles as in Section 1, i.e., the range of the differentials of j_n-J_n- and j-J-holomorphic maps, respectively. The $\mathscr{E}_n$, as subbundles of $T\,\mathrm{pr}^*TM$, together with their complex structures $J_n^{(1)}$ converge in C^∞ to $(\mathscr{E}, J^{(1)})$, in the sense that suitable local real basis fields $X_1^n, \ldots, X_{m+1}^n, J_n^{(1)}X_1^n, \ldots, J_n^{(1)}X_{m+1}^n$ converge in C^∞ to some local basis fields $X_1, \ldots, X_{m+1}, J^{(1)}X_1, \ldots, J^{(1)}X_{m+1}$.

Now construct almost complex structures $J_n^{(1)}$ and $J^{(1)}$ on $[(G, j_n), (M, J_n)] \simeq \mathrm{pr}^*TM$ and $[(G, j), (M, J)] \simeq \mathrm{pr}^*TM$ as in Section 1. To that end choose a fixed subbundle $\mathscr{H}$ of $T\,\mathrm{pr}^*TM$ complementary to the vertical bundle and a sequence (h_n) of Hermitian metrics on $G \times M$ converging in C^∞ to h. Next, extend the complex structure of $\mathscr{E}_n \to \mathrm{pr}^*TM$ to an almost complex structure $J_n^{(1)}$ of pr^*TM using $\mathscr{H}$ and h_n. Then $J_n^{(1)}$ converges in C^∞ to $J^{(1)}$. Using this, Proposition 3.1 generalizes to

Proposition 4.1 (generalized Weierstraß theorem). *Assume (S, j) is a Riemann surface without boundary and (M, J) an almost complex manifold. Let (j_n) and (J_n) be sequences of complex structures on S and almost complex structures on M converging in C^∞ to j and J, respectively. Then any C^0-convergent sequence $f_n: (S, j_n) \to (M, J_n)$, $n \geq 1$, of j_n-J_n-holomorphic maps converges even in C^∞ to a j-J-holomorphic map.*

Proof. We may assume that S is an open subset of $\mathbb{R}^2$ and then the proof of Proposition 3.1 can be copied step by step. The estimates of $\|Tf_n\|$ follow by the remarks at the beginning of this section. The ranges of the 1-jets are now pr^*TM via the above identification. Note that a 1-form β on V satisfying (i) in the above proof also satisfies $2d\beta(v, J_n^{(1)}v) \geq \mu_n^{(1)}(v, v)$ for each $v \in TV$ and each sufficiently large n. $\qquad\square$

Chapter IV

Hyperbolic surfaces

This chapter describes the pairs of pants decomposition and the thick-thin decomposition of oriented complete hyperbolic surfaces of finite area. Using the pairs of pants decomposition, one gets, roughly speaking, a parametrization of the space of hyperbolic structures on such a surface which coincides with the space of its complex structures. The thick-thin decomposition gives a classification of the thin parts of a hyperbolic surface, which are the components of small injectivity radius. Basic references for this chapter are [Ab] and [Bu].

1. Hexagons

We denote by $\mathbb{H}$ the upper half-space model of the hyperbolic plane and we recall Remark I.4.5 which says that we do not strictly distinguish between parametrized and unparametrized geodesic segments, in abuse of notation and terminology.

A subset $C \subset \mathbb{H}$ is called *convex* if for any two points $p, q \in C$ the geodesic segment connecting p to q is contained in C.

A *compact right angled hexagon $G \subset \mathbb{H}$* is a convex compact subset of $\mathbb{H}$ where the boundary consists of six consecutive geodesic segments $a_1, b_1, a_2, b_2, a_3, b_3$ with all its interior angles at the corners, that are the endpoints of the *sides* $a_1, \ldots, b_3$, equal to $\pi/2$ (see Figure 4). We parametrize each of the sides with constant speed 1 such that the vectors $i\dot{a}_1, \ldots, i\dot{b}_3$ point into the interior of G. With respect to this parametrization, the interior angle at some corner p is π minus the angle between the tangent vectors of the adjacent sides in $T_p\mathbb{H}$. In rough terms, a *degenerate right angled hexagon* is a right angled hexagon G in $\mathbb{H}$ with one or more of the sides b_1, b_2 or b_3 degenerate to a point in $\mathbb{H}(\infty)$. More precisely, it is a closed convex subset G of $\mathbb{H}$ with the following properties and with a degenerate side in the sense explained below:

(i) The topological boundary of G in $\overline{\mathbb{H}}$ is the union of six sets $a_1, b_1, a_2, b_2, a_3, b_3$ where each a_v is a geodesic segment and each b_v is a compact geodesic segment or a single point at infinity. We call the sets $a_1, \ldots, b_3$ the *sides* of G.

(ii) Each geodesic segment c in $\{a_1, \ldots, b_3\}$ is parametrized with constant speed 1 such that $i\dot{c}$ points into the interior of G for any interior point t of the parameter interval. We call the endpoints of c *corners* of G.

(iii) The sides $a_1, b_1, a_2, b_2, a_3, b_3$ are consecutive in the obvious sense.

(iv) The interior angles at the corners contained in $\mathbb{H}$ are equal to $\pi/2$.

A side which is a point at infinity we also call a *degenerate side* or a *side of length zero*.

Convention. By *hexagon* we always mean a right angled hexagon, compact or degenerate, unless otherwise stated.

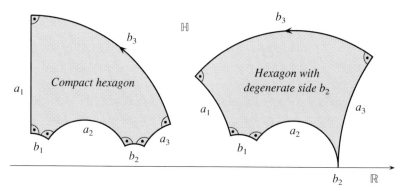

Figure 4.

Proposition 1.1. *Given $\ell_1, \ell_2, \ell_3 \in [0, \infty)$ there is up to orientation preserving isometry a unique hexagon G with consecutive sides $a_1, b_1, a_2, b_2, a_3, b_3$ satisfying $\ell(b_v) = \ell_v$ for $v = 1, 2, 3$, where $\ell(b_v)$ denotes the length of the side b_v.*

In the proof of this proposition and also in the sequel we implicitly use the following

Lemma 1.2.

(i)   *Given $p \in \mathbb{H}$ and a geodesic $c \subset \mathbb{H}$ there is a unique point $q =: \pi_c(p) \in c$ with $d(p, q) = d(p, c) := \inf \{ d(p, q') \mid q' \in c \}$. The geodesic segment connecting p to q intersects c orthogonally. The resulting map π_c extends to a continuous map $\pi_c \colon \overline{\mathbb{H}} \to c \subset \overline{\mathbb{H}}$.*

(ii) *Given two geodesics $c_1, c_2 \subset \mathbb{H}$ with $\inf \{ d(q_1, q_2) \mid q_v \in c_v \} =: d(c_1, c_2) > 0$ there is a unique geodesic segment c connecting c_1 and c_2 with $\ell(c) = d(c_1, c_2)$. This segment c intersects both c_1 and c_2 orthogonally.*

Proof. To prove (i) we may assume that $c = i\mathbb{R}_+$ and $p \notin c$. By completeness of the metric there is a $q \in c$ with $d(p, q) = d(p, c)$. Let c_1 be the geodesic segment from p to q. The reflection ρ at $i\mathbb{R}_+$, $\rho(z) = -\bar{z}$, is an isometry of $\mathbb{H}$. Thus $c_1 \cup \rho(c_1)$ is the shortest path connecting p and $\rho(p)$ for otherwise $d(p, c) < (1/2)\ell(c_1 \cup \rho(c_1)) = \ell(c_1) = d(p, q)$. Hence $c_1 \cup \rho(c_1)$ is a geodesic segment and it intersects $c = i\mathbb{R}_+$ orthogonally in q. This proves uniqueness of $q =: \pi_c(p)$. It follows from the construction that π_c is continuous on $\mathbb{H}$ and extends to a continuous map on $\overline{\mathbb{H}}$.

For the proof of (ii) we may assume that $c_1 = i\mathbb{R}_+$ and that c_2 has both of its endpoints in $\mathbb{R}_+ \subset \mathbb{H}(\infty)$. For $r \geq 0$ consider the set $A_r := \{ z \in \mathbb{H} \mid \operatorname{Re}(z) \geq 0, d(z, c_1) = r \}$. With c_1 also A_r is invariant under the isometries $T_l \colon z \mapsto e^l z, l \in \mathbb{R}$. Hence A_r is a Euclidean ray of the form $\mathbb{R}_+ w$ for $w \in A_r$. This implies that $A_{d(c_1, c_2)}$ is tangent to c_2 in some unique intersection point p and that the geodesic segment c from p to $\pi_{c_1}(p)$ is the unique path with $\ell(c) = d(c_1, c_2)$. It intersects c_1 orthogonally, and by symmetry also c_2. $\qquad\square$

Proof of Proposition 1.1. To begin with we give an explicit construction of a hexagon with prescribed lengths $\ell(b_v)$. If $\ell_1 = \ell_2 = \ell_3 = 0$ the convex set $G \subset \mathbb{H}$ bounded by $a_1 = i\mathbb{R}_+, a_2 = 1 + e^{i(0,\pi)}$ and $a_3 = 2 + i\mathbb{R}_+$ is a hexagon with $\ell(b_1) = \ell(b_2) = \ell(b_3) = 0$. We can now assume that at least one of the ℓ_v, say ℓ_1, is positive. Consider the geodesics α_1, α_2 and the geodesic segment b_1 with $\ell(b_1) = \ell_1$ as in Figure 5. The geodesic α_1 is the imaginary axis.

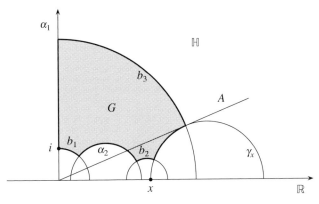

Figure 5.

Let A be the set of all points z in $\mathbb{H}$ with $d(z, \alpha_1) = \ell_3$ and $\mathrm{Re}(z) \geq 0$. From the last proof we know that A is a Euclidean ray. Assume first that $\ell_3 \neq 0$ and hence $A \neq \alpha_1$. For $x \geq \sup \mathrm{Re}(\alpha_2)$ consider the geodesic γ_x intersecting A tangentially and with one of its endpoints on $\mathbb{H}(\infty) = \mathbb{R} \cup \{\infty\}$ equal to x and the other greater than x. Choose x such that $d(\alpha_2, \gamma_x) = \ell_2$ and define $\alpha_3 = \gamma_x$. If $\ell_2 > 0$ let b_2 be the shortest geodesic segment connecting α_2 and α_3. Then $\ell(b_2) = d(\alpha_2, \alpha_3) = \ell_2$. If $\ell_2 = 0$ define $b_2 = \{x\}$. By construction $d(\alpha_3, \alpha_1) = \ell_3$. Let b_3 be the shortest geodesic segment connecting α_3 and α_1. Then $\ell(b_3) = \ell_3$. If $\ell_3 = 0$ then $A = \alpha_1$. In this case consider the geodesics $\gamma_x = x + i\mathbb{R}_+$ instead and argue as before. Then $b_3 = \{\infty\}$. So in any case we have defined the sides b_1, b_2, b_3 and hence G.

It remains to prove uniqueness. If G is a hexagon with three degenerate sides b_1, b_2, b_3 there is an isometry $\phi \in \mathrm{Iso}^+(\mathbb{H})$ with $\phi(b_1) = 0$, $\phi(b_2) = 1$ and $\phi(b_3) = \infty$. Otherwise note that everything but the choice of the oriented side b_1 in the above construction is unique. By transitivity of $\mathrm{Iso}^+(\mathbb{H})$ on unit tangent vectors this proves the uniqueness statement. $\qquad\square$

Proposition 1.3. *The area of any right angled hexagon is π.*

Before proving this proposition we compute the area of triangles. Three points in $x, y, z \in \overline{\mathbb{H}}$, not belonging to a single geodesic, define a triangle in $\mathbb{H}$, namely the smallest closed convex subset Δ of $\mathbb{H}$ with $x, y, z \in \overline{\Delta} \subset \overline{\mathbb{H}}$. We define the interior angle of a triangle at a corner at infinity to be 0.

Lemma 1.4. *The area of a triangle Δ in $\mathbb{H}$ with interior angles α, β and γ is*

$$\mathscr{A}(\Delta) = \pi - \alpha - \beta - \gamma .$$

Sketch of proof. The volume form of $\mathbb{H}$ is $y^{-2}dx \wedge dy = d(y^{-1}dx)$. First suppose that $\gamma = 0$. Then we may assume that the corresponding corner is ∞. Applying Stokes' theorem one gets $\mathscr{A}(\Delta) = \pi - \alpha - \beta$. To prove the claim in general note that any triangle Δ with all its corners in $\mathbb{H}$ is the closure of a set-theoretic difference $\Delta = \Delta_1 \setminus \Delta_2$ of two triangles each having one corner in the boundary at infinity, as shown in Figure 6 below.   □

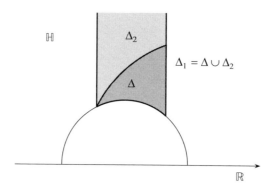

Figure 6.

Proof of Proposition 1.3. Decompose a given hexagon into triangles and apply the lemma.   □

Now consider again a hexagon $G \subset \mathbb{H}$ with consecutive sides $a_1, \ldots, b_3$. Let α_v be the geodesic in $\mathbb{H}$ with $a_v \subset \alpha_v$ and extend the parametrization of a_v to a parametrization of α_v. Define $q_v := \pi_{\alpha_{v+1}}(\alpha_v(-\infty))$ and $p_v := \pi_{\alpha_v}(\alpha_{v+1}(\infty))$ where the indices are taken modulo 3 (see Figure 7). The points q_{v-1}, p_v are contained in a_v and $p_v \geq q_{v-1}$ with respect to the order on a_v induced by its parametrization. Moreover, if $\ell(b_v) > 0$ then $d(p_v, b_v) = d(q_v, b_v) =: w(b_v)$ by symmetry.

We define a distinguished collar $\mathscr{C}(b_v)$ around each of the sides b_v in G. If $\ell(b_v) > 0$ then let

$$\mathscr{C}(b_v) := \left\{ z \in G \mid d(z, b_v) \leq w(b_v) \right\}$$

and if $\ell(b_v) = 0$ then define

$$\mathscr{C}(b_v) := G \cap \overline{\mathrm{HB}}(b_v, p_v),$$

where $\overline{\mathrm{HB}}(b_v, p_v)$ denotes the closed horoball centred at b_v whose boundary contains p_v. We note that the points p_v, q_v are contained in the boundary of $\mathscr{C}(b_v)$.

Lemma 1.5. *The interiors of the collars $\mathscr{C}(b_v)$ of a hexagon G are pairwise disjoint.*

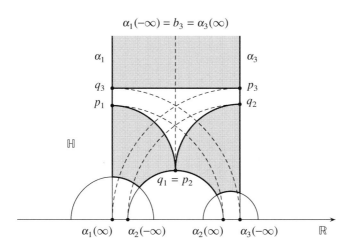

Figure 7.

Proof. In case $\ell(b_1), \ell(b_2) > 0$ the geodesic segment a_2 is the shortest path connecting b_1 and b_2. Since $p_2 \geq q_1$, the interiors of their collars are disjoint. If $b_3 = \{\infty\}$ and $\ell(b_2) > 0$ the segment a_3 contains the unique shortest path connecting b_2 and the horosphere $\{z \mid \mathrm{Im}(z) = \mathrm{Im}(p_3)\}$. Again, since $p_3 \geq q_2$, this proves the claim in this situation. In case $\ell(b_1) = \ell(b_2) = 0$ and $b_1 = \{\infty\}$ the claim follows from the above description of $\mathscr{C}(b_2)$ again using $p_2 \geq q_1$. $\square$

Lemma 1.6. *Suppose that G is a hexagon as above.*
(i)   *If $\ell(b_v) > 0$ then $\sinh \ell(b_v) \cdot \sinh w(b_v) = 1$.*
(ii) *The distance between the points p_v and q_v satisfies $d(p_v, q_v) > \operatorname{arcsinh}(1)$.*

Proof. This proof is illustrated in Figure 8. In order to prove (i) we may assume without loss of generality that b_v is the geodesic segment from $ie^{\ell(b_v)}$ to i. It is a consequence of the proof of Lemma 1.2 that

$$\mathscr{C}(b_v) = \left\{ re^{it} \mid \theta \leq t \leq \tfrac{\pi}{2}, 1 \leq r \leq y \right\}$$

for a suitable $\theta \in (0, \pi/2)$ and $y := \left| ie^{\ell(b_v)} \right|$. The geodesic segment joining $1 = \alpha_{v+1}(\infty)$ and $p_v = \pi_{\alpha_v}(1)$ is a segment of a Euclidean circle. Denote by x its centre on $\mathbb{R} \subset \mathbb{H}(\infty)$. Considering the right angled Euclidean triangle with corners $0, x$ and p_v, we deduce that

$$\tan \frac{\theta}{2} = \frac{1 - \cos \theta}{\sin \theta} = \frac{1 - \frac{y}{x}}{\frac{x-1}{x}} = \frac{y - 1}{y + 1}$$

since $y^2 + (x-1)^2 = x^2$. The hyperbolic length of the geodesic segment $c\colon t \mapsto y e^{it}$, $t \in [\theta, \pi/2]$, is the width $w(b_v)$, that is

$$w(b_v) = \int_\theta^{\frac{\pi}{2}} \|\dot c(t)\|\, dt = \int_\theta^{\frac{\pi}{2}} \frac{dt}{\sin t} = -\log \tan \frac{\theta}{2} = \log \frac{y+1}{y-1}.$$

Now it is easy to compute $\sinh w(b_v)$, namely

$$2 \sinh w(b_v) = \frac{y+1}{y-1} - \frac{y-1}{y+1}.$$

Since $2 \sinh \ell(b_v) = y - y^{-1}$, we get immediately that $\sinh w(b_v) \cdot \sinh \ell(b_v) = 1$.

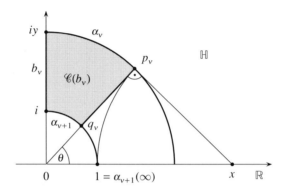

Figure 8.

It remains to estimate the distance between the points p_v and q_v. To begin with assume that $\ell(b_v) > 0$. In view of Figure 8 and using (I.4.3) we obtain that

$$\begin{aligned}
\sinh\left(\tfrac{1}{2} d(p_v, q_v)\right) &= \sinh\left(\tfrac{1}{2} d(e^{i\theta}, y e^{i\theta})\right) \\
&= \frac{y-1}{2(y \sin^2 \theta)^{1/2}} \\
&= \frac{y-1}{2\sqrt{y}} \cdot \frac{x}{x-1} = \frac{1}{2\sqrt{y}} \cdot \frac{y^2+1}{y+1} \\
&\geq \frac{1}{2}.
\end{aligned} \tag{1.1}$$

In order to prove the last estimate note that $y \geq 1$ and equality holds in (1.1) for $y = 1$. By differentiation with respect to y one proves the inequality. This implies the desired estimate since $2\,\mathrm{arcsinh}(1/2) > \mathrm{arcsinh}(1)$.

If $\ell(b_v) = 0$ then we see from Figure 7 that $d(p_v, q_v) = d(y + iy, iy)$ for some $y > 0$ and thus

$$\sinh\left(\tfrac{1}{2} d(p_v, q_v)\right) = \frac{y}{2(y^2)^{1/2}} = \frac{1}{2}.$$

Consequently, the estimate also holds for degenerate boundary components, and this finishes the proof.   □

2. Building hyperbolic surfaces from pairs of pants

Definition. Assume S is an orientable surface with all its boundary components diffeomorphic to the circle S^1. A complete Riemannian metric h on S is called a *hyperbolic structure* on S if each $s \in S$ has some neighbourhood isometric to some open subset of $\{ z \in \mathbb{H} \mid \operatorname{Re}(z) \geq 0 \}$. We call such a pair (S, h) a *hyperbolic surface*.

Remark. The boundary components of a hyperbolic surface are closed geodesics in the sense that each boundary component is diffeomorphic to S^1 and each boundary point has some neighbourhood isometric to some subset U of $\mathbb{H}$ where $U \setminus \operatorname{int}(U)$ is a geodesic segment in $\mathbb{H}$. Equivalently one could thus define a hyperbolic surface to be a surface together with a complete Riemannian metric h of constant curvature -1 such that all the boundary components are closed geodesics.

Lemma 2.1. *Let (S, h) be a hyperbolic surface. The Riemannian universal covering space $\tilde{S}$ of (S, h) is (isometric to) a closed subset of $\mathbb{H}$ where each boundary component is a geodesic in $\mathbb{H}$ and the deck transformations are restrictions of elements in* $\operatorname{Iso}^+ (\mathbb{H})$.

Remark. The *Riemannian universal covering space* of a Riemannian manifold is the universal covering space together with the lifted Riemannian metric.

Proof. Let γ denote a boundary component of S. A sufficiently small tubular neighbourhood of γ in S is isometric to the corresponding tubular neighbourhood of the boundary of $Z_\gamma := \langle T_{\ell(\gamma)} \rangle \backslash \{ z \in \mathbb{H} \mid \operatorname{Re}(z) \geq 0 \}$. For each boundary component γ of S glue Z_γ to S along γ such that the two parts fit together metrically. In this way a complete orientable surface S' without boundary is obtained, which is locally isometric to $\mathbb{H}$ and contains S as a closed subset with geodesic boundary. Hence the Riemannian universal covering space of S' is $\mathbb{H}$ and clearly, it contains $\tilde{S}$ as a closed subset of $\mathbb{H}$ with geodesic boundary.   □

Let G be a hexagon with consecutive sides $a_1, b_1, a_2, b_2, a_3, b_3$ where $0 \leq k \leq 3$ of the sides b_1, b_2, b_3 are degenerate and let G' be a copy of G with corresponding sides $a'_1, \ldots, b'_3$ and corresponding parametrizations. Identifying, for $v = 1, 2, 3$ and t in the parameter interval of a_v, the point $a_v(t)$ with $a'_v(t)$ we obtain a surface Y diffeomorphic to a 2-sphere with $3 - k$ pairwise disjoint open discs and k points removed (see also Figure 9). The complex structures i of $G \subset \mathbb{H}$ and $-i$ of $G' \subset \mathbb{H}$ fit together and define a complex structure on Y. Since we glued along geodesic segments, Y inherits a Hermitian metric of constant curvature -1 from G and G'. The surface Y together with its complex structure and Hermitian metric is called a *pair of pants*. Making the same construction with the closures $\overline{G}$ and $\overline{G}'$ in $\overline{\mathbb{H}}$, we obtain a compact topological space $\overline{Y}$, a compactification of Y.

Definition. Let Y be a pair of pants. We call every point $\gamma \in \overline{Y} \setminus Y$ a *degenerate boundary component* or a *boundary component of length* $\ell(\gamma) := 0$.

In Section 1 we defined distinguished collars $\mathscr{C}(b_v)$ around the sides b_v of a hexagon G. These collars together with their copies in G' define *distinguished collars* $\mathscr{C}(\gamma)$ around the boundary components γ of the pair of pants Y obtained by gluing G and G' together. Before summarizing their properties in the next proposition, we recall that T_l denotes the hyperbolic isometry $T_l\colon z \mapsto e^l z$ and P the parabolic isometry $P\colon z \mapsto z + 1$ of $\mathbb{H}$.

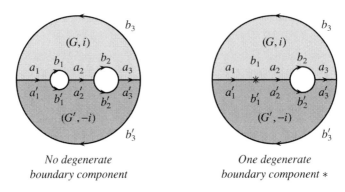

No degenerate
boundary component

One degenerate
boundary component $*$

Figure 9.

Proposition 2.2. *Let Y be a pair of pants. Then the following holds.*

(i) *Assume γ is a closed boundary geodesic of Y with length l. Then γ has a closed tubular neighbourhood $\mathscr{C}(\gamma)$ isometric to*

$$\langle T_l \rangle \Big\backslash \Big\{ z \in \mathbb{H} \,\big|\, \operatorname{Re} z \leq 0,\ \sinh(d(z, i\mathbb{R}_+)) \cdot \sinh(\tfrac{1}{2}l) \leq 1 \Big\}.$$

(ii) *Suppose that γ is a degenerate boundary component of Y. Then there is a closed subset $\mathscr{C}(\gamma)$ isometric to*

$$\langle P \rangle \Big\backslash \Big\{ z \in \mathbb{H} \,\big|\, \operatorname{Im} z \geq \tfrac{1}{2} \Big\}$$

such that $\mathscr{C}(\gamma) \cup \{\gamma\}$ is a neighbourhood of γ in $\overline{Y}$.

(iii) *The interiors of the distinguished collars of Y are pairwise disjoint.*

Proof. We obtain (i) directly from Lemma 1.6.(i). In order to prove (ii) we may assume that Y results from gluing hexagons G and G' where $b_3 = \infty$ is the degenerate side of G corresponding to γ (refer also to Figure 7). One easily checks that the length of the boundary of $\langle P \rangle \backslash \{ z \in \mathbb{H} \mid \operatorname{Im} z \geq 1/2 \}$ is equal to two times the length of the horosphere segment joining q_3 and p_3, and this shows (ii). Part (iii) follows immediately from Lemma 1.5. $\qquad\square$

Definition. A surface together with a Riemannian metric which is isometric to

$$\langle P\rangle\backslash\left\{z\in\mathbb{H}\mid \operatorname{Im}z\geq r\right\}$$

for some $r>0$ is called a *cusp*. If the surface is isometric to

$$\langle P\rangle\backslash\left\{z\in\mathbb{H}\mid \operatorname{Im}z\geq \tfrac{1}{2}\right\}$$

we call it a *standard cusp*.

Remark 2.3. We note that the area of a pair of pants is 2π by Proposition 1.3. In particular the area of a cusp is finite.

Corollary 2.4. *Each pair of pants is a hyperbolic surface.*

Proof. It remains only to show that the Hermitian metric of each non-compact pair of pants is complete. This follows from the proposition since cusps are complete. □

Corollary 2.5. *The compactification $\overline{Y}$ of a pair of pants Y can be turned into a Riemann surface, in a unique way, such that the inclusion $Y\hookrightarrow\overline{Y}$ becomes a holomorphic embedding.*

Proof. Cusps are conformally equivalent to $\overline{D}\setminus\{0\}$ (see Section I.5). □

Definition. A *marking* of a pair of pants Y is a numbering $(\gamma_1,\gamma_2,\gamma_3)$ of its (possibly degenerate) boundary components. Together with a marking, Y is called a *marked* pair of pants.

On a marked pair of pants $(Y,(\gamma_1,\gamma_2,\gamma_3))$ we fix three canonical points $y_1,y_2,y_3\in\overline{Y}$ uniquely determined by the following properties, where the indices are taken modulo 3:

(i) $y_\nu=\gamma_\nu$ if $\gamma_\nu\in\overline{Y}\setminus Y$;

(ii) if γ_ν and $\gamma_{\nu-1}$ are boundary geodesics let $y_\nu\in\gamma_\nu$ be the starting point of the shortest geodesic segment in Y connecting γ_ν to $\gamma_{\nu-1}$;

(iii) if γ_ν is a boundary geodesic and $\gamma_{\nu-1}\in\overline{Y}\setminus Y$ a degenerate boundary component let $y_\nu\in\gamma_\nu$ be the starting point of the geodesic segment $\alpha\colon[0,\infty)\to Y$ which intersects γ_ν in $\alpha(0)$ orthogonally and satisfies $\alpha(t)\to\gamma_{\nu-1}$ as $t\to\infty$.

Remark 2.6. From the construction of the pairs of pants the existence of the geodesics in (ii) and (iii) is clear. After passing to the universal covering space, uniqueness in (ii) and (iii) follows from Lemma 1.2 and Lemma 1.4, respectively. Note that the y_ν are the forward endpoints of the $a_\nu\subset G$ and this shows that the decomposition of a pair of pants into hexagons is unique.

If γ_ν is a boundary geodesic of Y, we parametrize γ_ν with constant speed $\|\dot\gamma_\nu\|\equiv\ell(\gamma_\nu)$ and $\gamma_\nu(0)=y_\nu$. We choose its orientation so that $i\dot\gamma_\nu$ points into the interior of Y.

From the geometry of hexagons in $\mathbb{H}$ (see Section 1) the following proposition can be deduced immediately.

Proposition 2.7. *Up to isometry there exists exactly one pair of pants with prescribed lengths of its boundary components.*

Proof. Immediate from Proposition 1.1. ☐

We now consider a family $(Y^i, (\gamma_1^i, \gamma_2^i, \gamma_3^i))_{i=1,\ldots,N}$ of N marked pairs of pants and let $\gamma^1, \tilde{\gamma}^1, \ldots, \gamma^n, \tilde{\gamma}^n$ be a family of pairwise different boundary components in $\{ \gamma_k^i \mid i = 1, \ldots, N; k = 1, 2, 3 \}$. Their parametrizations shall be the distinguished ones from above. We assume that $\ell(\gamma^v) = \ell(\tilde{\gamma}^v)$ for $v = 1, \ldots, n$ and we pick $\alpha_1, \ldots, \alpha_n \in [0, 1]$. For $v = 1, \ldots, n$ and $t \in [0, 1]$ we identify

$$\gamma^v(t) = \begin{cases} \tilde{\gamma}^v(\alpha_v - t), & \text{if } t \le \alpha_v \\ \tilde{\gamma}^v(\alpha_v - t + 1), & \text{if } t > \alpha_v. \end{cases}$$

In this way we obtain an oriented surface S. As before, when gluing hexagons together, the hyperbolic structures of the pairs of pants fit together and define a hyperbolic structure on S. This defines a complex structure on the oriented surface S compatible with the hyperbolic structure. The parameters $\alpha_1, \ldots, \alpha_n$ are called *twist-parameters*, and we they can be viewed as elements of the circle $\mathbb{Z} \backslash \mathbb{R}$.

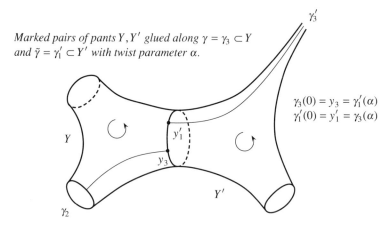

Marked pairs of pants Y, Y' glued along $\gamma = \gamma_3 \subset Y$ and $\tilde{\gamma} = \gamma_1' \subset Y'$ with twist parameter α.

$$\gamma_3(0) = y_3 = \gamma_1'(\alpha)$$
$$\gamma_1'(0) = y_1' = \gamma_3(\alpha)$$

Figure 10.

Suppose that S is connected and closed. Then $N =: 2g - 2$ is even. The integer $g \ge 2$ determines S up to diffeomorphism among the connected, closed orientable surfaces and it is called the *genus* of S. Of course, one cannot build spheres and tori by gluing together pairs of pants. To these surfaces of exceptional type one associates genus zero and one, respectively. One can show that each closed orientable surface is diffeomorphic to a surface of genus g for some unique $g \ge 0$ (see for instance [Hi] or [Ma]).

Assume now that S is connected but not necessarily closed and let k denote the number of its isometrically embedded standard cusps and m the number of its non-degenerate boundary components. Then there exists a unique integer $g \geq 0$ such that S is diffeomorphic to a closed orientable surface of genus g with m pairwise disjoint open discs and k points removed. Then g is called the *genus* of S and (g, m, k) its *signature*. The pairs of pants $Y^1, \ldots, Y^N$ are isometrically embedded in S.

For non-negative integers g, k, m with $2g + m + k \geq 3$ a hyperbolic surface of signature (g, m, k) is obtained by suitably gluing together $2g - 2 + m + k$ pairs of pants as above. Since the area of each hexagon is π, this shows the following

Proposition 2.8. *The area of a hyperbolic surface of signature (g, m, k) depends only on (g, m, k) and is given by $2\pi(2g - 2 + m + k)$.*

3. Pairs of pants decomposition

The aim of this section is to show that each hyperbolic surface *diffeomorphic* to some "model surface", as constructed in the last section, can be decomposed into pairs of pants and thus is in fact *isometric* to some "model surface".

Recall that T_l denotes the hyperbolic isometry $T_l \colon z \mapsto e^l z$ and P the parabolic isometry $P \colon z \mapsto z + 1$ of $\mathbb{H}$.

Lemma 3.1. *Assume $\eta \colon \mathbb{R} \to \mathbb{H}$ is a continuous curve satisfying $\eta(\tau + 1) = \sigma_0(\eta(\tau))$ for each $\tau \in \mathbb{R}$ and some fixed point free $\sigma_0 \in \mathrm{Iso}^+(\mathbb{H})$. Then $\eta(\mathbb{R}) \cap \sigma(\eta(\mathbb{R})) \neq \varnothing$ for each $\sigma \in \mathrm{Iso}^+(\mathbb{H})$ having exactly the same fixed points at infinity as σ_0.*

Proof. Assume $c \colon \mathbb{R} \to \mathbb{H}$ is a geodesic or a horosphere parametrized with constant speed $u > 0$. We recall from Remark I.4.12 that the map $\mathbb{R}^2 \to \mathbb{H}$, given by

$$(t, s) \mapsto \exp_{c(t)}\left(s \frac{i\dot{c}(t)}{u}\right)$$

is a diffeomorphism and that its inverse map are so-called Fermi coordinates with respect to c.

Using Fermi coordinates with respect to the axis of σ_0 if σ_0 is hyperbolic, or with respect to some σ_0-invariant horosphere if σ_0 is parabolic, the lemma reduces to proving the following

Claim. Let e_1, e_2 denote the standard basis of $\mathbb{R}^2$. If $\eta \colon \mathbb{R} \to \mathbb{R}^2$ is a continuous curve satisfying $\eta(\tau + 1) = e_1 + \eta(\tau)$ for each $\tau \in \mathbb{R}$ then $(re_1 + \eta(\mathbb{R})) \cap \eta(\mathbb{R}) \neq \varnothing$ for any $r \in \mathbb{R}$.

Indeed, in corresponding Fermi coordinates (t, s), the isometries $\sigma \in \mathrm{Iso}^+(\mathbb{H})$ having the same fixed points as σ_0 are given by $(t, s) \mapsto (t + r, s)$ since distance tubes around the axis or σ_0-invariant horospheres are preserved under each such σ.

In order to prove the claim, let η_1, η_2 denote the components of η. By 1-periodicity of η_2 there exist $\tau_* < 0$ and $\tau^* > 1$ such that

$$\eta_2(\tau_*) = \min \eta_2(\mathbb{R}) \quad \text{and} \quad \eta_2(\tau^*) = \max \eta_2(\mathbb{R}).$$

Furthermore, observe that for any given points r and x in $\mathbb{R}$,

$$(\{x\} \times \mathbb{R}) \cap (re_1 + \eta(\mathbb{R})) \neq \emptyset.$$

Thus, an intermediate value argument shows that $\eta([\tau_*, \tau^*])$ has a non-empty intersection with $re_1 + \eta(\mathbb{R})$. □

Before continuing we introduce some terminology. We consider the universal covering $\tilde{S} \xrightarrow{p} S$ of a surface S and denote by Σ the group of deck transformations. Then for any subset $V \subset \tilde{S}$ we set

$$\Sigma_V := \left\{ \sigma \in \Sigma \mid \sigma(V) = V \right\}$$

which is a subgroup of Σ. A subset $V \subset \tilde{S}$ is called *precisely Σ-invariant* if $\sigma(V) \cap V = \emptyset$ for each $\sigma \in \Sigma \setminus \Sigma_V$. If $V \subset \tilde{S}$ is precisely Σ-invariant then

$$\Sigma_V \backslash^V \hookrightarrow \Sigma \backslash^{\tilde{S}} = S, \quad \Sigma_V v \mapsto \Sigma v$$

is an embedding. Notice also that for any path-connected subset $U \subset S$ each component of $p^{-1}(U)$ is precisely Σ-invariant.

Proposition 3.2. *Assume S is a surface diffeomorphic to a closed orientable surface with finitely many disjoint open discs removed and let S^* be a surface obtained from S by removing finitely many points. Assume h is a hyperbolic structure on S^* with finite area. Let $c \subset S^*$ be a simple closed loop which is homotopically non-trivial in S^*. Then exactly one of the following assertions holds:*

(i) *There exists a unique closed geodesic γ in the free homotopy class of c. Moreover, γ is simple closed and if $\gamma \cap c = \emptyset$ the loops γ and c bound an annulus in S^*.*

(ii) *The loop c bounds a disc in S containing exactly one point of $S \setminus S^*$ in its interior.*

Before proving the proposition, we describe the ends of such a hyperbolic surface.

Lemma 3.3. *Let S^* be as in the proposition and let $U \subset S$ be a neighbourhood of some point in $S \setminus S^*$. Then $U \cap S^* \subset S^*$ contains a cusp.*

Proof of the lemma. We denote by $\tilde{S}^* \xrightarrow{p} S^*$ the Riemannian universal covering and by Σ the group of deck transformations. In view of Lemma 2.1 we consider $\tilde{S}^*$ as subset of $\mathbb{H}$ and Σ as subgroup of $\mathrm{Iso}^+(\mathbb{H})$.

Let Δ be a closed neighbourhood of some point in $S \setminus S^*$ such that $\Delta^* := \Delta \cap S^*$ is a semi-open annulus. Pick a connected component $V \subset \tilde{S}^*$ of $p^{-1}(\Delta^*)$. By the above,

$$\Delta^* \simeq \Sigma_V \backslash^V \subset \Sigma_V \backslash^{\mathbb{H}}$$

and thus Σ_V is an infinite cyclic subgroup of Σ. Moreover, since the area of S^* is finite and S^* is complete, Σ_V is generated by a parabolic isometry of $\mathbb{H}$ and Δ^* contains a closed subset isometric to some cusp. □

Remark. The following proof makes use of the Jordan-Schoenflies theorem which states that a simple closed C^0-loop in $\mathbb{R}^2$ bounds an open disc (see [Mo]).

Proof of the proposition. We denote by $[c]$ the free homotopy class of the loop c in S^* and define $\ell([c]) := \inf \{ \ell(c') \mid c' \in [c] \}$. First assume that $\ell([c]) > 0$. Then choose a sequence $(c_n)_{n \geq 1}$ in $[c]$ with

$$\lim_{n \to \infty} \ell(c_n) = \ell([c]).$$

By the previous lemma there exists a neighbourhood U of $S \setminus S^*$ with $c_n \cap U = \varnothing$ for each n. Assuming that each c_n is parametrized with constant speed on S^1, it follows from Arzelà-Ascoli's theorem that the sequence $(c_n : S^1 \to S^*)_{n \geq 1}$ has a C^1-convergent subsequence (compare Proposition B.3). Let $\gamma: S^1 \to S^*$ be a limit of such a convergent subsequence. Then $\gamma \in [c]$ and $\ell(\gamma) = \ell([c])$. Thus γ is a closed geodesic since it is locally distance minimizing.

Now choose a homotopy $\Phi: S^1 \times [0, 1] \to S^*$ from $\gamma = \Phi(\cdot, 0)$ to $c = \Phi(\cdot, 1)$. Let ρ and p denote the universal covering maps as in the diagram

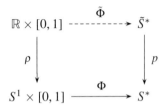

and let $\tilde{\Phi}$ be a lift of $\Phi \circ \rho$. Again, denote the deck transformation group of $\tilde{S}^* \to S^*$ by Σ and consider $\tilde{S}^*$ as subset of $\mathbb{H}$ and Σ as subgroup of $\mathrm{Iso}^+(\mathbb{H})$. Then $\tilde{\gamma} := \tilde{\Phi}(\cdot, 0)$ and $\tilde{c} := \tilde{\Phi}(\cdot, 1)$ are lifts of γ and c, respectively, to $\tilde{S}^*$, and $d(\tilde{\gamma}(t), \tilde{c}(t)) < C$ for some constant C and each $t \in \mathbb{R}$.

The points of distance smaller than C from the imaginary axis in $\mathbb{H}$ are contained in some cone bounded by rays $\mathbb{R}_+ w, \mathbb{R}_+(-\overline{w})$, for some $w \in \mathbb{H}$ (see the proof of Lemma 1.2). This implies that

$$\lim_{t \to \pm\infty} \tilde{c}(t) = \tilde{\gamma}(\pm\infty) \qquad (3.1)$$

which proves the uniqueness statement in (i).

Since c is simple closed, $\tilde{c}$, as a point set, is precisely Σ-invariant. The closed geodesic γ has no transversal self-intersection. Otherwise there would exist a $\sigma \in \Sigma$ with $\sigma(\tilde{\gamma}) \cap \tilde{\gamma}$ consisting of exactly one point. By (3.1) neither $\sigma(\tilde{c}) \cap \tilde{c} = \varnothing$ nor $\sigma(\tilde{c}) \cap \tilde{c} = \tilde{c}$ in contradiction to $\tilde{c}$ being precisely Σ-invariant. In order to show that γ is simple closed, one has to show that $\Sigma_{\tilde{c}} = \Sigma_{\tilde{\gamma}}$ with

$$\Sigma_{\tilde{c}} = \Big\{ \sigma \in \Sigma \mid \sigma(\tilde{c}) = \tilde{c} \text{ as point sets} \Big\}.$$

From (3.1) it follows that $\Sigma_{\tilde{c}} \subset \Sigma_{\tilde{\gamma}}$. Now pick any $\sigma \in \Sigma_{\tilde{\gamma}}$. Lemma 3.1 implies that $\sigma(\tilde{c}) \cap \tilde{c} \neq \varnothing$. Hence $\sigma(\tilde{c}) \cap \tilde{c} = \tilde{c}$ since $\tilde{c}$ is precisely Σ-invariant and thus $\sigma \in \Sigma_{\tilde{c}}$.

If $\gamma \cap c = \varnothing$ the lifts $\tilde{\gamma}$ and $\tilde{c}$ bound a strip W in $\tilde{S}^*$ whose interior is diffeomorphic to $\mathbb{R} \times (0, 1)$ by the Jordan-Schoenflies theorem, as illustrated in Figure 11. Moreover, W is precisely Σ-invariant with $\Sigma_W = \Sigma_{\tilde{c}}$ being infinitely cyclic. This implies that the annulus $\Sigma_W \backslash W$ is isometrically embedded in S^* and bounded by $\gamma \cup c$.

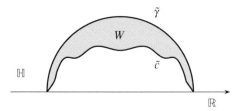

Figure 11.

Now suppose that $\ell([c]) = 0$. Since $[c]$ is homotopically non-trivial in S^*, for any closed neighbourhood U of $S \setminus S^*$ in S there exists some $c' \in [c]$ with $c' \subset U$. By the previous lemma there exists a cusp $\Delta^* \subset S^*$ containing arbitrarily small $c'' \in [c]$. After possibly making the cusp smaller, we may assume that $c \cap \Delta^* = \varnothing$. By the above, we can write $\Delta^* = \Sigma_V \backslash V$ where V is some closed horoball in $\tilde{S}^* \subset \mathbb{H}$. Without loss of generality, assume that the centre of the horoball V is ∞. Since each loop in Δ^* is retractable into the boundary $\partial \Delta^*$, there exists some loop $\gamma \in [c]$ contained in $\partial \Delta^*$.

Then choose a homotopy Φ from c to γ and consider a lift $\tilde{\Phi}$ as above. We may assume that $\tilde{\Phi}(\mathbb{R} \times \{0\}) = \partial V$. Again the Jordan-Schoenflies theorem implies that the corresponding lifts $\tilde{c}$ and $\tilde{\gamma}$ bound a strip W diffeomorphic to $\mathbb{R} \times (0, 1)$ (see Figure 12). Indeed, the closure of the set $\tilde{c} \cup \tilde{\gamma}$ in $\overline{\mathbb{H}}$ is the simple closed loop $\tilde{c} \cup \tilde{\gamma} \cup \{\infty\}$ since $d(\tilde{c}(t), \tilde{\gamma}(t)) < C$ for some constant C and each $t \in \mathbb{R}$.

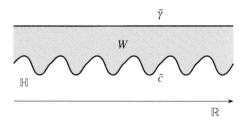

Figure 12.

Now consider a non-trivial isometry $\sigma \in \mathrm{Iso}^+(\mathbb{H})$ fixing $\tilde{c}$ as point set. Consequently, such σ also fixes the closure of $\tilde{c}$ in $\overline{\mathbb{H}}$ and thus fixes the centre ∞ of V. We claim that this is the only fixed point of σ in $\overline{\mathbb{H}}$. Otherwise σ would be hyperbolic and without loss of generality σ would also fix 0. Thus $\sigma = T_l \colon z \mapsto e^l z$ for some $l \in \mathbb{R} \setminus \{0\}$, contradicting the fact that σ fixes $\tilde{c}$ as a point set. Hence σ is parabolic with fixed point ∞

and thus fixes the horosphere $\tilde{\gamma}$ centred at ∞ as a point set. This implies that $\Sigma_{\tilde{c}} \subset \Sigma_{\tilde{\gamma}}$ and arguing as before using Lemma 3.1, it follows that $\Sigma_{\tilde{c}} = \Sigma_{\tilde{\gamma}}$. This shows that γ is simple closed in S^* and that (ii) holds under the assumption $\ell([c]) = 0$.

Finally note that (i) and (ii) cannot hold simultaneously since the previous arguments show that any simple closed geodesic in S^* is not homotopically equivalent to the boundary of a cusp. $\qquad\square$

The Jordan-Schoenflies theorem immediately implies the following remark about simple closed loops in hyperbolic surfaces.

Remark 3.4. *Any homotopically trivial simple closed loop in a hyperbolic surface bounds an open disc.*

Proof. Let $\tilde{S} \xrightarrow{p} S$ be the universal covering of a given hyperbolic surface S and Σ the group of deck transformations. Then any lift of a homotopically trivial simple closed loop $\gamma \subset S$ to $\tilde{S}$ is simple closed and bounds a disc $\Delta \subset \mathbb{H}$ by the Jordan-Schoenflies theorem. It is evident that $\Delta \subset \tilde{S}$. Since γ is simple closed, $\partial\Delta$ and hence also Δ are precisely Σ-invariant. And since any homeomorphism from Δ to itself has a fixed point, we see that $\Sigma_\Delta = \{\mathbf{1}\}$. Consequently, $p|_\Delta$ is a homeomorphism. $\qquad\square$

Lemma 3.5. *Let S be a surface diffeomorphic to a pair of pants and h a hyperbolic structure on S of finite area. Then (S, h) is isometric to some pair of pants.*

Proof. There is an integer $0 \leq m \leq 3$ such that S is diffeomorphic to S^2 with m open discs $\Delta_1, \ldots, \Delta_m$ and $3 - m$ points $p_{m+1}, \ldots, p_3$ removed. For each point p_ν we choose a closed disc $\overline{\Delta}_\nu$ in S containing p_ν in its interior Δ_ν such that the closed discs $\overline{\Delta}_1, \overline{\Delta}_2$ and $\overline{\Delta}_3$ are pairwise disjoint. Denote by γ_1, γ_2 and γ_3 the boundaries of Δ_1, Δ_2 and Δ_3, respectively.

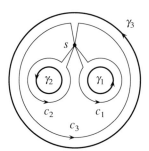

Figure 13.

Let $\tilde{S} \to S$ be the Riemannian universal covering of S and Σ its deck transformation group. Again, view $\tilde{S}$ as a closed subset of $\mathbb{H}$ and Σ as a subgroup of $\mathrm{Iso}^+(\mathbb{H})$. After choosing suitable orientations for the loops γ_ν, each loop corresponds to a conjugacy class of an isometry $\sigma_\nu \in \Sigma$ satisfying $\sigma_1 \sigma_2 = \sigma_3$ as we can see from the topology of

S as follows. Choose some $s \in S \setminus \partial S$ and identify Σ in a natural manner with the fundamental group $\pi_1(S, s)$. Let σ_1, σ_2 and σ_3 be the isometries corresponding to the elements in $\pi_1(S, s)$ represented by loops c_1, c_2 and c_3, respectively, as in Figure 13.

For $v = 1, \ldots, m$ the isometry σ_v leaves some geodesic in $\tilde{S} \subset \mathbb{H}$ projecting to γ_v invariant. Hence σ_v is a hyperbolic isometry with displacement $\ell(\gamma_v)$. Arguing as in the final part of the last proof, it follows that σ_v is parabolic for $v = m+1, \ldots, 3$ since the area of S is finite. Hence, in view of Lemma I.4.10, we know $|\mathrm{tr}(\sigma_v)|$ from $\ell(\gamma_v)$ for $v = 1, 2, 3$. We claim that the group Σ can be determined from the data $|\mathrm{tr}(\sigma_1)|, |\mathrm{tr}(\sigma_2)|, |\mathrm{tr}(\sigma_3)|$ and the relation $\sigma_1 \sigma_2 = \sigma_3$ up to conjugacy in $\mathrm{Iso}^+(\mathbb{H})$ and some non-relevant reflection. This will finish the proof. Indeed, using Proposition 2.7, pick a pair of pants Y such that the lengths of its boundary components coincide with the lengths of the boundary components of S. The previous considerations and the claim show that Y is isometric to S.

It remains to prove the claim. First assume that σ_1 is hyperbolic. After conjugation of Σ in $\mathrm{Iso}^+(\mathbb{H}) \simeq \mathrm{PSL}(2, \mathbb{R})$ we may assume that

$$\sigma_1 = \begin{bmatrix} a & 0 \\ 0 & a^{-1} \end{bmatrix}$$

for some $a > 1$ by Lemma I.4.9 and (I.4.6). Moreover, after conjugation with a suitable T_l, we can additionally assume that $\sigma_2(\infty) \in \{\pm 1\}$. Suppose first that $\sigma_2(\infty) = 1$. Then σ_2 and thus σ_3 are of the form

$$\sigma_2 = \begin{bmatrix} b & c - b^{-1} \\ b & c \end{bmatrix} \quad \text{and} \quad \sigma_3 = \sigma_1 \sigma_2 = \begin{bmatrix} ab & * \\ * & a^{-1}c \end{bmatrix}$$

for some b, c. Now $a > 1$ is determined by $|\mathrm{tr}\,\sigma_1|$ and then, given $a > 1$, σ_2 and σ_3 are determined by $|\mathrm{tr}(\sigma_2)|$ and $|\mathrm{tr}(\sigma_3)|$. The solution for $\sigma_2(\infty) = -1$ is obtained from the previous one by conjugation with the orientation reversing isometry $z \mapsto -\bar{z}$.

If all the isometries σ_1, σ_2 and σ_3 are parabolic we may assume that the corresponding fixed points of σ_1 and σ_2 are ∞ and 0, respectively. Moreover, conjugation with a suitable T_l yields that $\sigma_1 \in \{P, P^{-1}\}$ by Lemma I.4.9. Suppose that $\sigma_1 = P$. Again this determines $\sigma_1, \sigma_2, \sigma_3$ uniquely. To that end note that $|\mathrm{tr}(\sigma_v)| = 2$ for each v by Lemma I.4.10. Thus

$$\sigma_1 = \begin{bmatrix} 1 & 1 \\ 0 & 1 \end{bmatrix}, \quad \sigma_2 = \begin{bmatrix} 1 & 0 \\ b & 1 \end{bmatrix} \quad \text{and} \quad \sigma_3 = \begin{bmatrix} 1+b & 1 \\ b & 1 \end{bmatrix}$$

and the unique solution is $b = -4$. Again, conjugation of this solution with the reflection $z \mapsto -\bar{z}$ gives the solution for $\sigma_1 = P^{-1}$. $\square$

Suppose that (S, h) is a surface together with some complete Riemannian metric and $\mathcal{L} = (c_1, \ldots, c_N)$ is a family of pairwise disjoint smooth simple closed loops in $S \setminus \partial S$. Here we mean by a smooth simple closed loop a smoothly embedded S^1. Consider $S'_\mathcal{L} := S \setminus \bigcup_{i=1}^N c_i$ together with the Riemannian metric $h'_\mathcal{L}$ induced from h by restriction and let $d'_\mathcal{L}$ be the distance function of $(S'_\mathcal{L}, h'_\mathcal{L})$. The metric completion $(S_\mathcal{L}, d_\mathcal{L})$

of $(S'_{\mathscr{L}}, d'_{\mathscr{L}})$ carries a natural differentiable structure and $d_{\mathscr{L}}$ is induced by a Riemann-ian metric $h_{\mathscr{L}}$ on $S_{\mathscr{L}}$. Then $(S_{\mathscr{L}}, h_{\mathscr{L}})$ is called the surface obtained by *cutting* (S, h) *open along* $\mathscr{L}$. Note that, up to diffeomorphism, $S_{\mathscr{L}}$ does not depend on the choice of the complete metric g.

Now assume (S, h) is a hyperbolic surface. A family $\mathscr{L}$ of pairwise disjoint simple closed geodesics *decomposes* (S, h) *into pairs of pants* if each component of $(S_{\mathscr{L}}, h_{\mathscr{L}})$ is a pair of pants. Of course, one recovers (S, h) by regluing the components of $(S_{\mathscr{L}}, h_{\mathscr{L}})$ in the right fashion.

Proposition 3.6. *Let S be a surface obtained from a closed orientable surface of genus g by removing m open discs and k points. Assume h is a hyperbolic structure on S with finite area. Then (S, h) can be decomposed into pairs of pants, i.e., it is (isometric to) a hyperbolic surface of signature (g, m, k).*

Proof. There is a family of simple closed loops $c_1, \ldots, c_N$, $N = 3g - 3 + m + k$, which decomposes S into pieces each *diffeomorphic* to a pair of pants. Let γ_v be the geodesic loop in the free homotopy class of c_v determined by Proposition 3.2. Arguing as in the proof of Proposition 3.2, especially (3.1), and using the fact that $c_1, \ldots, c_N$ are pairwise disjoint, it follows that $\gamma_1, \ldots, \gamma_N$ are also pairwise disjoint. Thus $\gamma_1, \ldots, \gamma_N$ decompose (S, h) into pairs of pants. $\square$

The next theorem due to L. Bers guarantees the existence of decompositions with a priori bounds for the lengths of the boundaries of the pairs of pants.

Theorem 3.7 (Bers). *Let (S, h) be a hyperbolic surface of signature (g, m, k). Then there exists a constant C, depending only on the length of the boundary of S and the signature (g, m, k) of S, and a decomposition of S into pairs of pants such that the lengths of each of their boundary components are bounded by C.*

The proof of this theorem follows from

Lemma 3.8. *Let (S, h) be as in the theorem. If S is not a pair of pants then there exists a simple closed geodesic in S disjoint from ∂S of length smaller than $\ell(\partial S) + 5\mathscr{A}(S)$.*

Proof of the lemma. First suppose that S is closed. Let γ be a shortest closed geodesic in S. Then for any $p \in \gamma$ the ball $B_{\ell(\gamma)/2}(p) \subset S$ is isometric to an $\ell(\gamma)/2$-ball in $\mathbb{H}$. Otherwise injrad$(S, p) < \ell(\gamma)/2$ and thus there would exist a simple closed geodesic loop c through p with $\ell(c) = 2$ injrad$(S, p) < \ell(\gamma)$. Using Proposition 3.2, we would obtain a simple closed geodesic γ' in S with $\ell(\gamma') \leq \ell(c) < \ell(\gamma)$, a contradiction. We claim that

$$\mathscr{A}(S) \geq \mathscr{A}(B_{\ell(\gamma)/2}(p)) \geq \tfrac{\pi}{4}\ell(\gamma)^2 . \tag{3.2}$$

Indeed, the area of an R-ball B_R in $\mathbb{H}$ is greater than the area of an R-ball in the Euclidean plane. This follows from the description of the hyperbolic metric in polar coordinates in Remark I.4.11,

$$\mathscr{A}(B_R) = \int_0^R \int_0^{2\pi} \sinh(r)\, d\theta\, dr > \int_0^R \int_0^{2\pi} r\, d\theta\, dr = \pi R^2 .$$

The inequality (3.2) proves the claim in case S is closed, since $\mathcal{A}(S) \geq \sqrt{\mathcal{A}(S)} > 1$.

Now assume that S is not closed. Consider around each boundary geodesic γ of S the open collars

$$\mathcal{C}'(\gamma, r) = \left\{ s \in S \mid d(s, \gamma) < r \right\}$$

for $r > 0$ so small that $\mathcal{C}'(\gamma, r)$ is contained in the distinguished collar $\mathcal{C}(\gamma)$. If $\ell(\gamma) < 1$, then we see from part (ii) of Lemma 1.6 that there exists an $r(\gamma) > 0$ such that $\mathcal{C}'(\gamma) := \mathcal{C}'(\gamma, r(\gamma)) \subset \mathcal{C}(\gamma)$ and $\ell(\{ s \in S \mid d(s, \gamma) = r(\gamma) \}) = 1$. For a degenerate boundary component γ of S let $\mathcal{C}_1(\gamma)$ be the cusp contained in the standard cusp $\mathcal{C}(\gamma)$ with $\ell(\partial \mathcal{C}_1(\gamma)) = 1$. To that end observe that the length of the boundary of the standard cusp is greater than 1. Let $\mathcal{C}'(\gamma)$ be the interior of $\mathcal{C}_1(\gamma)$ in S. Now define

$$S' := S \setminus \bigcup_{\gamma \in \Gamma} \mathcal{C}'(\gamma)$$

where Γ is the set of those possibly degenerate boundary component of S with their lengths being smaller than 1. Observe that $S' \subset S$ is a compact surface with boundary and each boundary component of S' has length greater or equal to 1.

By $\mathrm{tb}(r) := \{ p \in S' \mid d(p, \partial S') \leq r \}$ we denote the closed r-tube around $\partial S'$ in S'. For $r > 0$ small enough $\mathrm{tb}(r)$ is diffeomorphic to $\partial S' \times [0, 1]$. We put

$$r_0 := \sup \left\{ r > 0 \mid \mathrm{tb}(r) \text{ is diffeomorphic to } \partial S' \times [0, 1] \right\}.$$

Let γ be any boundary component of S'. The area of the component of $\mathrm{tb}(r)$, $r < r_0$, containing γ is greater than the area of a corresponding collar in the Euclidean cylinder with circumference equal to $\ell(\gamma)$. Again, this can be readily verified using Fermi coordinates as described in Remark I.4.12 and we obtain that

$$\mathcal{A}(S) \geq \mathcal{A}(S') \geq \mathcal{A}(\mathrm{tb}(r)) \geq r_0 \ell(\partial S') \geq r_0 \tag{3.3}$$

since $\ell(\partial S') \geq 1$ by the definition of S'. By definition of r_0 there exist components γ_1 and γ_2 of $\partial S'$ and a point $p \in S'$ with

$$d(p, \gamma_1) = d(p, \gamma_2) = r_0.$$

Hence, there is a piecewise geodesic segment β of length $2r_0$ containing p with one endpoint on γ_1 and the other one on γ_2. By minimality of r_0 the segment β is in fact a geodesic segment. We distinguish now two cases.

CASE 1. $\gamma_1 \neq \gamma_2$. Parametrize γ_1 and γ_2 on $[0, 1]$ such that their endpoints lie on β and such that $i\dot{\gamma}_\nu$ points into the interior of S' where i denotes a compatible complex structure on S. Orient β such that it goes from γ_1 to γ_2 and denote by β^- the geodesic β with reversed orientation. The loop $\gamma_1 \beta \gamma_2 \beta^-$ is homotopic to a simple closed loop c in S' which does not intersect $\partial S'$ (see Figure 14).

Moreover, we can assume that c is contained in an arbitrarily small neighbourhood of $\gamma_1 \beta \gamma_2 \beta^-$ in S' and that $\ell(c)$ is arbitrarily close to $\ell(\gamma_1 \beta \gamma_2 \beta^-)$. By Remark 3.4 the loop

c is homotopically non-trivial for otherwise S would be diffeomorphic to a cylinder. We claim that there exists a simple closed geodesic γ in S freely homotopic to $\gamma_1 \beta \gamma_2 \beta^-$. Indeed, otherwise c would bound a punctured disc in S and S would be a pair of pants by Proposition 3.2. Again by Proposition 3.2, the closed geodesic γ is not a boundary component of ∂S since S is not a pair of pants.

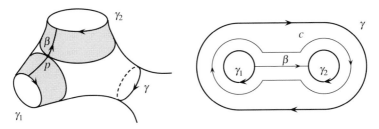

Figure 14.

Using (3.3), the length of γ can be estimated in the following way:

$$\ell(\gamma) \leq \ell(\gamma_1 \beta \gamma_2 \beta^-) = \ell(\gamma_1) + \ell(\gamma_2) + 4r_0$$
$$\leq \ell(\partial S) + 1 + 1 + 4\mathcal{A}(S)$$
$$< \ell(\partial S) + 5\mathcal{A}(S)$$

since $\mathcal{A}(S) \geq 2\pi$. Consequently, γ has the properties claimed in the lemma.

CASE 2. $\gamma_1 = \gamma_2$. By construction β has two different endpoints on γ_1. They decompose γ_1 into two segments γ', γ''. Now parametrize γ', γ'' and β such that $\gamma'\beta^-$ and $\gamma''\beta$ are simple closed loops (see Figure 15).

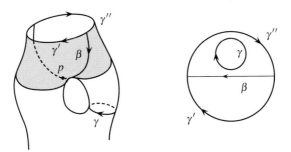

Figure 15.

We claim that each of the loops $\gamma'\beta^-$ and $\gamma''\beta$ is homotopically non-trivial and not homotopic to γ_1. If this were not true, then some lift of β and a suitable lift of γ' or γ'' to the universal covering space would have the same endpoints contradicting the fact that there is a unique geodesic segment connecting two given points in $\mathbb{H}$.

Observe that at least one of the simple closed loops $\gamma'\beta^-$ and $\gamma''\beta$ is homotopic to a simple closed geodesic in S which is not a boundary component of S for otherwise S would be a pair of pants by Proposition 3.2.

The length of the simple closed geodesic γ can be estimated as in Case 1.   □

Proof of Theorem 3.7. Let (S, h) be a hyperbolic surface of signature (g, m, k). The area of S is $\mathcal{A}(S) = 2\pi(2g - 2 + m + k)$ by Proposition 2.8 and depends only on the signature of S. Using the lemma, choose a simple closed geodesic γ of length smaller than $\ell(\partial S) + 5\mathcal{A}(S)$ and cut S open along γ. In this way proceed with each component of the resulting surface which is not a pair of pants. Each of these components again satisfies the assumption of the lemma and the length of its boundary is bounded in terms of $\ell(\partial S)$ and $\mathcal{A}(S)$. The procedure stops after at most $3g - 3 + m + k$ steps.   □

The next proposition shows that Bers' theorem applies especially to closed Riemann surfaces with finitely many points removed and equipped with their Poincaré metric. This is an essential ingredient of Gromov's proof of his compactness theorem for J-holomorphic curves.

Proposition 3.9. *Assume S is a closed Riemann surface of genus $g \geq 0$ and let $s_1, \ldots, s_k$ be k different points in S. We further assume that the Riemann surface $S^* := S \setminus \{s_1, \ldots, s_k\}$ is of non-exceptional type. Then the area of S^* is finite with respect to its Poincaré metric h. Thus (S^*, h) is a hyperbolic surface of signature $(g, 0, k)$.*

Proof. For each $s \in \{s_1, \ldots, s_k\}$ choose an open neighbourhood U of s in S conformally equivalent to the open unit disc $D \subset \mathbb{C}$ such that $U^* := U \setminus \{s\} \subset S^*$. Denote by $p: \mathbb{H} \to (S^*, h)$ and $q: \mathbb{H} \to (U^*, h')$ the corresponding Riemannian universal coverings where h' denotes the Poincaré metric of U^*. Now lift the inclusion $U^* \hookrightarrow S^*$ to a holomorphic map $\mathbb{H} \to \mathbb{H}$ of the covering spaces. The (Gromov-)Schwarz lemma implies that its differential is universally bounded. And, since (U^*, h') is a parabolic cylinder, this yields that it has finite area with respect to h. Hence the area of (S^*, h) is finite.   □

4. Thick-thin decomposition

This section shows what the thin components of a hyperbolic surface of signature $(g, 0, k)$ look like.

Lemma 4.1. *Let S be a hyperbolic surface of signature $(g, 0, k)$. Then the simple closed geodesics in S of lengths smaller than $2\operatorname{arcsinh}(1)$ are pairwise disjoint. In particular, there are only finitely many of them and their number is bounded by $3g - 3 + k$.*

Proof. Suppose γ is a simple closed geodesic in S of length smaller than $2\operatorname{arcsinh}(1)$. Proposition 2.2 shows that γ has some neighbourhood $W(\gamma)$ isometric to

$$\langle T_{\ell(\gamma)} \rangle \setminus \left\{ z \in \mathbb{H} \ \middle| \ \sinh(d(z, i\mathbb{R}_+)) \cdot \sinh(\tfrac{1}{2}\ell(\gamma)) < 1 \right\}.$$

and that the arcsinh(1)-distance tube around the simple closed geodesic γ, that is

$$\text{tb}(\gamma) := \left\{ s \in S \mid d(s, \gamma) < \text{arcsinh}(1) \right\},$$

is contained in $W(\gamma)$. Now assume that γ' is another simple closed geodesic in S of length smaller than $2\,\text{arcsinh}(1)$ which intersects γ. This implies that $\gamma' \subset \text{tb}(\gamma)$ and thus $\gamma' = \gamma$ since γ is the only simple closed geodesic contained in $W(\gamma)$.

Note that any two different simple closed geodesics are non-homotopic. Thus, the number of simple closed geodesics of length smaller than $2\,\text{arcsinh}(1)$ can be estimated from above by the number of simple closed geodesics needed to decompose S into pairs of pants, and that is $3g - 3 + k$. $\qquad\square$

Proposition 4.2. *Let S be a hyperbolic surface of signature $(g, 0, k)$ and $U \subset S$ a component of $\{ s \in S \mid \text{injrad}(S, s) < \text{arcsinh}(1) \}$. Then either*

(i) *U contains a simple closed geodesic γ of S of length $l = \ell(\gamma) < 2\,\text{arcsinh}(1)$ and is isometric to*

$$\langle T_l \rangle \backslash \left\{ z \in \mathbb{H} \mid d(z, e^l z) < 2\,\text{arcsinh}(1) \right\}$$

or

(ii) *the closure of U in S is a standard cusp and hence isometric to*

$$\langle P \rangle \backslash \left\{ z \in \mathbb{H} \mid \text{Im}\, z \geq \tfrac{1}{2} \right\}.$$

Moreover, the injectivity radius at a point $s \in U$ of S equals the injectivity radius at a corresponding point in $\langle T_l \rangle \backslash \mathbb{H}$ or $\langle P \rangle \backslash \mathbb{H}$ with respect to (i) or (ii).

Proof. Let γ be a simple closed geodesic in S of length smaller than $2\,\text{arcsinh}(1)$. Denote by $W(\gamma)$ the distinguished neighbourhood of γ as in the proof of the lemma. Observe from part (ii) of Lemma 1.6 that $W(\gamma)$ contains some open neighbourhood $U(\gamma)$ of γ isometric to

$$\langle T_{\ell(\gamma)} \rangle \backslash \left\{ z \in \mathbb{H} \mid d(z, e^{\ell(\gamma)} z) < 2\,\text{arcsinh}(1) \right\}.$$

Let $\gamma_1, \ldots, \gamma_n$ be the simple closed geodesics of S which have length smaller than $2\,\text{arcsinh}(1)$ and let $V_1, \ldots, V_k$ be the standard cusps of S. The open sets $U(\gamma_1), \ldots, U(\gamma_n)$ and $V_1, \ldots, V_k$ are pairwise disjoint since the sets $W(\gamma_1), \ldots, W(\gamma_n)$ and $V_1, \ldots, V_k$ are so by Proposition 2.2. Clearly, the sets $U(\gamma_1), \ldots, U(\gamma_n)$ and $V_1, \ldots, V_k$ are components of

$$\left\{ s \in S \mid \text{injrad}(S, s) < \text{arcsinh}(1) \right\}.$$

It remains to show that in fact these are all components.

Let $p \colon \mathbb{H} \to S$ be the Riemannian universal covering and Σ the group of deck transformations. Let $s \in S$ be any point with $\text{injrad}(S, s) < \text{arcsinh}(1)$ and choose $z \in p^{-1}(\{s\})$.

By Lemma I.4.8 there exists some $\sigma \in \Sigma$ with $d(z, \sigma(z)) = 2 \operatorname{injrad}(S, s)$. If σ is hyperbolic its axis projects to a simple closed geodesic $\gamma \in \{\gamma_1, \ldots, \gamma_n\}$ and consequently, $s \in U(\gamma)$. In case that σ is parabolic consider the geodesic loop $c := p \circ \tilde{c}$ where $\tilde{c}$ is the geodesic segment connecting z to $\sigma(z)$. Then Proposition 3.2 implies that c bounds a connected, closed subset of S which contains a cusp. The corresponding standard cusp C is

$$\langle \sigma \rangle \backslash \left\{ z \in \mathbb{H} \mid d(z, \sigma(z)) \leq 2 \operatorname{arcsinh}(1) \right\} \hookrightarrow S$$

and thus $s \subset C$. $\qquad \square$

5. Compactness properties of hyperbolic structures

In this section compactness properties of the space of hyperbolic structures on a given surface S with given signature are investigated. In this context, we use a theorem of H. Whitney [Wh] in order to extend certain functions, defined on closed subsets of some $\mathbb{R}^m$, to the ambient space. A good reference for that are §§1 and 2 in Chapter VI of [St].

The space of hyperbolic structures on an orientable surface S, up to orientation preserving diffeomorphisms, is called *Riemann space*. It is a quotient of the *Teichmüller space* of S. These spaces are studied in detail in the literature (see for instance [Ab] and for compact S also [Bu]). However, for our purposes it is not necessary to introduce these spaces explicitly.

Consider a sequence $(S_n^*)_{n \geq 1}$ of hyperbolic surfaces of the same signature (g, m, k). Denote by S_n the Riemann surface obtained by the one-point compactification of each standard cusp of S_n^*. We suppose that the length of any closed geodesic of S_n^*, which is not a boundary component, is bounded from below by some positive constant $c > 0$ independent of n. Moreover, assume the length of any boundary component of S_n^* to be bounded from above by some constant independent of n.

Now decompose each hyperbolic surface S_n^* into pairs of pants $Y_n^1, \ldots, Y_n^N$, $N = 2g - 2 + m + k$. Using Bers' Theorem 3.7, we may assume that all lengths of boundary components of the pairs of pants are bounded from above, independent of n. The number of pairs of pants decompositions of a hyperbolic surface is bounded in terms of its signature. Thus, after passing to a subsequence of (S_n^*), again denoted by (S_n^*), the following assumption can be made. For each n there exists an orientation preserving diffeomorphism $\chi_n : S_1^* \to S_n^*$ with $\chi_n(Y_1^i) = Y_n^i$ for $i = 1, \ldots, N$. Here the Y_n^i should be viewed canonically as subsets of S_n^*.

Next, choose a marking for each of the pairs of pants $Y_1^1, \ldots, Y_1^N$ and then attach to $Y_n^1, \ldots, Y_n^N$, $n \geq 2$, the markings induced by χ_n. Let $l_{11}, \ldots, l_{1r}$, $r = 3g - 3 + k + m$, denote the lengths of the simple closed geodesics along which S_1^* has been cut open and $\alpha_{11}, \ldots, \alpha_{1r} \in \mathbb{Z} \backslash \mathbb{R}$ the corresponding twist-parameters. Let $L_{11}, \ldots, L_{1s}$, $s = m + k$, be the lengths of the (possibly degenerate) boundary components of S_1^*. Denote by $l_{n1}, \ldots, l_{nr}, \alpha_{n1}, \ldots, \alpha_{nr}, L_{n1}, \ldots, L_{ns}$ the parameters of S_n^* corresponding to those of S_1^* with respect to χ_n. These parameters are called *Fenchel-Nielsen parameters* for S_n^*.

By the assumption and the choice of the pairs of pants decompositions, the lengths $l_{n1}, \ldots, l_{nr}, L_{n1}, \ldots, L_{ns}$ are bounded uniformly, i.e., independently of n. Thus we can choose a further subsequence of (S_n^*) such that the Fenchel-Nielsen parameters of S_n^* converge as $n \to \infty$, say

$$l_{ni} \to l_i > 0, \quad \alpha_{ni} \to \alpha_i \quad \text{and} \quad L_{nj} \to L_j \tag{5.1}$$

for $i = 1, \ldots, r$ and $j = 1, \ldots, s$. Let $k' + k$, $k' \geq 0$, be the number of the L_j which are zero.

In the next step construct a hyperbolic surface S^* of signature $(g, m - k', k + k')$ with Fenchel-Nielsen parameters $l_1, \ldots, l_r, \alpha_1, \ldots, \alpha_r, L_1, \ldots, L_s$ from marked pairs of pants $Y^1, \ldots, Y^N$. The marked pairs of pants $Y^1, \ldots, Y^N$ are those corresponding to $Y_n^1, \ldots, Y_n^N$ with respect to (5.1) and they are glued together in the corresponding way. Let S be the Riemann surface obtained from S^* by the one-point compactification of each standard cusp.

Let $\gamma_1^1, \ldots, \gamma_1^m$ be the non-degenerate, $\gamma_1^{m+1}, \ldots, \gamma_1^{m+k}$ the degenerate boundary components of S_1^*. Denote by $\gamma_n^1 := \chi_n(\gamma_1^1), \ldots, \gamma_n^{m+k} := \chi_n(\gamma_1^{m+k})$ and $\gamma^1, \ldots, \gamma^{m+k}$ the corresponding boundary components of S_n^* and S^*, respectively. Moreover, we may choose the indices in such a way that $\gamma^1, \ldots, \gamma^{k'}, \gamma^{m+1}, \ldots, \gamma^{m+k}$ are the degenerate boundary components of the hyperbolic surface S^* of signature $(g, m - k', k + k')$.

Proposition 5.1. *Under the above assumptions there is a sequence $\varphi_n \colon S_n \to S, n \geq 1$, of continuous maps with the following properties:*

(i) $\varphi_n(\gamma_n^i) = \gamma^i$ *for $i = 1, \ldots, m + k$ and each n.*

(ii) *For each n the map φ_n restricted to the set $S_n \setminus \bigcup_{i=1}^{k'} \gamma_n^i$ is a diffeomorphism onto $S \setminus \bigcup_{i=1}^{k'} \gamma^i$. We denote by ψ_n its inverse map.*

(iii) *The pull back $\psi_n^* h_n$ of the hyperbolic structure h_n of S_n^* is a Riemannian metric on S^*. The sequence $(\psi_n^* h_n)$ converges in the C^∞-topology to the hyperbolic structure of S^*.*

(iv) *The sequence $(\psi_n^* j_n)$ of pull backs of the complex structures j_n of S_n converges in the C^∞-topology to the complex structure j of S restricted to $S \setminus \bigcup_{i=1}^{k'} \gamma^i$.*

Remark. This proposition yields a differential geometric description of the Deligne-Mumford compactification of the Riemann space of a closed oriented surface of genus ≥ 2 (refer to [DM]). Compare also Appendix C in [Tr] on the Mumford compactness theorem.

The rest of this section is devoted to the proof of this proposition according to the following strategy. We give a constructive argument to find a sequence of maps (φ_n) having properties (i)–(iii). This reduces to give a constructive proof of the three claims below. Afterwards we show that we can handle condition (iv) additionally.

Let $(Y_n, (\gamma_n^1, \gamma_n^2, \gamma_n^3))_{n \geq 1}$ be a sequence of marked pairs of pants of the same signature. We assume that the lengths of their boundary components converge as $n \to \infty$, namely

$$\lim_{n \to \infty} \ell(\gamma_n^\nu) =: L_\nu \quad \text{for } \nu = 1, 2, 3.$$

Let $(Y_\infty, (\gamma_\infty^1, \gamma_\infty^2, \gamma_\infty^3))$ be the marked pair of pants with $\ell(\gamma_\infty^v) = L_v$. Let h_n, $n \geq 1$, and h_∞ denote the hyperbolic structures of Y_n and Y_∞, respectively. Finally, fix some positive integer $r \geq 1$.

Claim 1. If $L_v > 0$ for $v = 1, 2, 3$ then there exists a sequence

$$\psi_n : Y_\infty \to Y_n, \qquad n \geq 1,$$

of marking preserving C^∞-diffeomorphisms such that the sequence $(\psi_n^* h_n)_{n \geq 1}$ converges in the C^{r-1}-topology to h_∞.

For a pair of pants Y, a boundary component γ of Y and $\delta \in (0, \operatorname{arcsinh} 1)$ we define $\mathscr{C}(\gamma, \delta) := \{ s \in \mathscr{C}(\gamma) \mid \operatorname{injrad}(Y, s) < \delta \} \subset Y$ where $\mathscr{C}(\gamma)$ is the distinguished collar as in Proposition 2.2. Let $I := \{ v \mid L_v = 0 \} \subset \{ 1, 2, 3 \}$.

Claim 2. For any $\delta \in (0, \operatorname{arcsinh} 1)$ there is a sequence

$$\psi_n : Y_\infty \to Y_n \setminus \bigcup_{v \in I} \gamma_n^v, \qquad n \geq 1,$$

of marking preserving C^∞-diffeomorphisms such that the sequence $(\psi_n^* h_n)_{n \geq 1}$ restricted to the subset

$$Y_\infty \setminus \bigcup_{v \in I} \mathscr{C}(\gamma_\infty^v, \delta)$$

converges in the C^{r-1}-topology to h_∞, restricted to the same subset.

Let $(\tilde{Y}_n, (\tilde{\gamma}_n^1, \tilde{\gamma}_n^2, \tilde{\gamma}_n^3))_{n \geq 1}$ be another sequence of marked pairs of pants of the same signature, with the lengths of their boundary components converging as $n \to \infty$. We distinguish the corresponding expressions in Claim 1 and Claim 2 for $\tilde{Y}_n$ by $\tilde{\ }$. Furthermore, assume that $\ell(\gamma_n^1) = \ell(\tilde{\gamma}_n^1)$ for each n and $0 < L_1 = \tilde{L}_1$. Let (α_n) be a converging sequence in $\mathbb{Z} \backslash \mathbb{R}$ with limit α_∞. Denote by X_n, $1 \leq n \leq \infty$, the hyperbolic surface obtained by gluing Y_n to $\tilde{Y}_n$ along γ_n^1 and $\tilde{\gamma}_n^1$ with twist-parameter α_n. Let H_n be the hyperbolic structure of X_n. We can view Y_n and $\tilde{Y}_n$ canonically as subsets of X_n. Denote their common boundary component by γ_n.

Claim 3. For each $\delta \in (0, \operatorname{arcsinh} 1)$ the maps ψ_n and $\tilde{\psi}_n$ from the previous claims can be perturbed in $\mathscr{C}(\gamma_\infty^1)$ and $\mathscr{C}(\tilde{\gamma}_\infty^1)$, respectively, such that they fit together to C^∞-diffeomorphisms

$$\Psi_n : X_\infty \to X_n \setminus \left(\bigcup_{v \in I} \gamma_n^v \cup \bigcup_{v \in \tilde{I}} \tilde{\gamma}_n^v \right), \qquad n \geq 1,$$

and that the sequence $(\Psi_n^* H_n)$ restricted to the subset

$$X_\infty \setminus \left(\bigcup_{v \in I} \mathscr{C}(\gamma_\infty^v, \delta) \cup \bigcup_{v \in \tilde{I}} \mathscr{C}(\tilde{\gamma}_\infty^v, \delta) \right)$$

converges in the C^{r-1}-topology to H_∞ restricted to the same subset as $n \to \infty$.

Proof of Claim 1. For $1 \leq n \leq \infty$ we obtain Y_n by gluing together a right angled hexagon $G_n \subset \mathbb{H}$ with a copy G_n' of itself. Let $a_1^n, b_1^n, a_2^n, b_2^n, a_3^n, b_3^n$ denote the sides of G_n. Now choose a parametrization of the sides slightly different from the distinguished one of Section 1. Namely, parametrize them with constant speed over the unit interval $[0,1]$ such that their orientation is the distinguished one as in Section 1. Then define $\varepsilon := (1/8) \min\{L_1, L_2, L_3\} > 0$. We may assume that $b_1^\infty(0) = b_1^n(0)$ and $\dot{b}_1^\infty(0)/\|\dot{b}_1^\infty(0)\| = \dot{b}_1^n(0)/\|\dot{b}_1^n(0)\|$ for each n. For $v = 1,2,3$ define $\mathcal{T}(a_v^\infty) := \{z \in G_\infty \mid d(z, a_v^\infty) \leq \varepsilon\}$. The sets $(\mathcal{T}(a_v^\infty))_{v=1,2,3}$ are obviously pairwise disjoint. By $\mathcal{G}_\infty$ we denote the subset

$$\mathcal{G}_\infty := \bigcup_{v=1}^{3} (\mathcal{T}(a_v^\infty) \cup b_v^\infty)$$

of G_∞ as shown in the following figure.

Figure 16.

Let $\xi_v^n \colon \mathbb{H} \to \mathbb{R}^2$ be Fermi coordinates with respect to a_v^n as in Remark I.4.12, or more precisely, with respect to the geodesic which extends the geodesic segment a_v^n.
For each sufficiently large n a canonical map

$$\chi_n \colon \bigcup_{v=1}^{3} \mathcal{T}(a_v^\infty) \to G_n$$

is given by $\xi_v^n \circ \chi_n \circ (\xi_v^\infty)^{-1}\big|_{[0,1] \times [0,\varepsilon]} := \mathrm{id}_{[0,1] \times [0,\varepsilon]}$ for $v = 1,2,3$. Taking the special choice of the G_n into consideration, it follows that the sides of G_n converge to the sides of G_∞, in the sense that their parametrizations converge in C^∞. Thus the maps χ_n can be extended to C^∞-maps $\chi_n \colon \mathcal{G}_\infty \to G_n \subset \mathbb{H}$ such that they converge in C^∞ to the inclusion $\mathcal{G}_\infty \hookrightarrow \mathbb{H}$ as $n \to \infty$. We understand this in the following sense. It is possible to extend the maps χ_n to C^∞-maps defined on an open neighbourhood of $\mathcal{G}_\infty$ in $\mathbb{H}$ in such a way that all their derivatives converge locally uniformly on $\mathcal{G}_\infty$.

With Whitney's extension theorem (see [St], Chapter VI.2, Theorem 4) the maps χ_n, $1 \leq n \leq \infty$, can be extended to C^∞-maps $\tilde{\chi}_n \colon G_\infty \to \mathbb{H}$ such that the $\tilde{\chi}_n$ converge in C^r to $\tilde{\chi}_\infty$. For all sufficiently large n the map $\psi_n := \tilde{\chi}_n - \tilde{\chi}_\infty + \mathrm{id}_{G_\infty} \colon G_\infty \to \mathbb{H}$ is well defined and also an extension of χ_n. Here $+, -$ are the pointwise operations in the ambient space $\mathbb{C}$. The sequence (ψ_n) converges in C^r to the inclusion $G_\infty \hookrightarrow \mathbb{H}$. Since G_∞

is compact, the maps ψ_n are C^∞-diffeomorphisms $\psi_n\colon G_\infty \to G_n$ for all sufficiently large n. We denote by λ the Poincaré metric of $\mathbb{H}$. Then $\psi_n^*(\lambda|_{G_n})$ converges to $\lambda|_{G_\infty}$ in C^{r-1}.

By construction, the ψ_n fit together with their copies $\psi_n'\colon G_\infty' \to G_n'$ and define a C^∞-diffeomorphism $Y_\infty \to Y_n$, again denoted by ψ_n. It follows that $\psi_n^* h_n$ converges to h_∞ in C^{r-1} as $n \to \infty$. $\square$

We note that a similar construction as in the last proof can be found in [FLP].

Proof of Claim 2. In case $I = \{\, v \mid L_v = 0\,\} \neq \varnothing$ the previous construction has to be modified slightly. Again, Y_n is obtained by gluing together two degenerate hexagons $G_n, G_n' \subset \mathbb{H}$. For $\delta \in (0, \operatorname{arcsinh} 1)$ let

$$G_n[\delta] := G_n \setminus \bigcup_{v \in I} \mathscr{C}(\gamma_n^v, \delta) \subset \mathbb{H}.$$

When building the set-theoretic difference, view G_n as a subset of Y_n. Denote the corresponding sides of the compact "hexagon" $G_n[\delta]$ by $a_1^n[\delta], \ldots, b_3^n[\delta]$ and observe that $a_v^n[\delta] \subset a_v^n$ for $v = 1, 2, 3$. Then parametrize each side of $G_n[\delta]$ with constant speed over the interval $[0, 1]$ with the same convention for the orientation of the sides as above. In the same way as above we define $\mathscr{G}_\infty[\delta]$ to be the boundary of $G_\infty[\delta]$ together with ε-collars around the sides $a_1^\infty[\delta], a_2^\infty[\delta]$ and $a_3^\infty[\delta]$, where $\varepsilon > 0$ is sufficiently small depending on $G_\infty[\delta]$.

For the "hexagons" $G_n[\delta/2]$ now construct the maps ψ_n in a similar fashion to that above. Here we have to be more careful since the boundary of $G_n[\delta]$ is not piecewise geodesic. The construction below is illustrated in Figure 17.

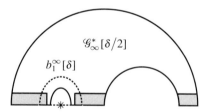

Figure 17.

First fix some $\delta \in (0, \operatorname{arcsinh} 1)$. We may clearly assume that the G_n have been chosen in such a way that $b_1^\infty[\delta/2](0) = b_1^n[\delta/2](0)$ and $\dot{b}_1^\infty[\delta/2](0)/\|\dot{b}_1^\infty[\delta/2](0)\| = \dot{b}_1^n[\delta/2](0)/\|\dot{b}_1^n[\delta/2](0)\|$. Thus the sides of $G_n[\delta/2]$ converge to the corresponding sides of $G_\infty[\delta/2]$ in C^∞ as $n \to \infty$. Now choose a small open neighbourhood U of $\bigcup_{v \in I} b_v^\infty[\delta/2]$ in $\mathscr{G}_\infty[\delta/2]$ which does not intersect $\mathscr{G}_\infty[\delta]$ and let

$$\mathscr{G}_\infty^*[\delta/2] := (\mathscr{G}_\infty[\delta/2] \setminus U) \cup \partial \mathscr{G}_\infty[\delta/2].$$

Then the maps $\chi_n \colon \mathscr{G}_\infty^*[\delta/2] \to G_n[\delta/2] \subset \mathbb{H}$ can be defined in the same way as in the proof of Claim 1 such that they converge in C^r to the inclusion $\mathscr{G}_\infty^*[\delta/2] \hookrightarrow \mathbb{H}$. Again, the χ_n can be extended to C^∞-diffeomorphisms $\psi_n \colon G_\infty[\delta/2] \to G_n[\delta/2] \subset \mathbb{H}$ converging in C^r to the inclusion $G_\infty[\delta/2] \hookrightarrow \mathbb{H}$.

The maps ψ_n and ψ_n', restricted to $G_\infty[\delta]$ and $G_\infty'[\delta]$, respectively, fit together by construction and define C^∞-maps

$$\widehat{\psi}_n \colon Y_\infty \setminus \bigcup_{v \in I} \mathscr{C}(\gamma_\infty^v, \delta) \to Y_n.$$

For all sufficiently large n the $\widehat{\psi}_n$ are embeddings and thus can be extended to diffeomorphisms

$$\psi_n \colon Y_\infty \to Y_n \setminus \bigcup_{v \in I} \gamma_n^v$$

having the desired properties. $\square$

Proof of Claim 3. For $1 \leq n \leq \infty$ consider the annulus $A_n := \mathscr{C}(\gamma_n^1) \cup \mathscr{C}(\tilde{\gamma}_n^1) \subset X_n$. Using Fermi coordinates with respect to the closed geodesics γ_n^1 in A_n, identify A_n with $\mathbb{Z}\backslash\mathbb{R} \times [-w_n, w_n]$ where w_n is the width of the collars $\mathscr{C}(\gamma_n^1)$ and $\mathscr{C}(\tilde{\gamma}_n^1)$. Here we point out that the parametrization of $\gamma_n^1 \subset Y_n$ is the one distinguished in Remark 2.6. For each sufficiently large n, by restriction of the maps ψ_n from Claim 1 & 2 we obtain maps

$$\tilde{\psi}_n \colon \mathbb{Z}\backslash\mathbb{R} \times [-w_\infty/2, 0] \to \mathbb{Z}\backslash\mathbb{R} \times [-w_n, 0] \subset \mathbb{Z}\backslash\mathbb{R} \times (-\infty, 0] \tag{5.2}$$

$$\psi_n \colon \mathbb{Z}\backslash\mathbb{R} \times [0, w_\infty/2] \to \mathbb{Z}\backslash\mathbb{R} \times [0, w_n] \subset \mathbb{Z}\backslash\mathbb{R} \times [0, \infty) \tag{5.3}$$

where the domain of $\tilde{\psi}_n$ and ψ_n corresponds to $\mathscr{C}(\tilde{\gamma}_n^1)$ and $\mathscr{C}(\gamma_n^1)$, respectively. We claim that these maps converge in C^r to the inclusions

$$\mathbb{Z}\backslash\mathbb{R} \times [-w_\infty/2, 0] \hookrightarrow \mathbb{Z}\backslash\mathbb{R} \times (-\infty, 0]$$
$$\mathbb{Z}\backslash\mathbb{R} \times [0, w_\infty/2] \hookrightarrow \mathbb{Z}\backslash\mathbb{R} \times [0, \infty).$$

Indeed, (ψ_n) converges in C^r by Claims 1 & 2 and $(\tilde{\psi}_n)$ converges since additionally the twist-parameters α_n converge.

Now lift the maps from (5.2) and (5.3) to maps $\tilde{f}_n$ and f_n between the universal covering spaces,

$$\tilde{f}_n \colon \mathbb{R} \times [-w_\infty/2, 0] \to \mathbb{R} \times (-\infty, 0]$$
$$f_n \colon \mathbb{R} \times [0, w_\infty/2] \to \mathbb{R} \times [0, \infty)$$

such that they converge in C^r to the inclusion. The maps $\tilde{f}_n$ and f_n can be perturbed on $\mathbb{R} \times [-w_\infty/4, 0]$ and $\mathbb{R} \times [0, w_\infty/4]$, respectively, so that they fit together to a sequence

$$F_n \colon \mathbb{R} \times [-w_\infty/2, w_\infty/2] \to \mathbb{R} \times (-\infty, \infty), \qquad n \geq 1,$$

of C^∞-maps which pass to maps $\mathbb{Z}\backslash\mathbb{R} \times [-w_\infty/2, w_\infty/2] \to \mathbb{Z}\backslash\mathbb{R} \times (-\infty, \infty)$ between the quotients and converge in C^r to the inclusion.

Indeed, pick a smooth map $\phi: [-w_\infty/2, w_\infty/2] \to [0, 1]$ with

$$\phi|_{[-w_\infty/2, -w_\infty/4]} \equiv 1, \quad \phi|_{[-w_\infty/8, w_\infty/8]} \equiv 0 \quad \text{and} \quad \phi|_{[w_\infty/4, w_\infty/2]} \equiv 1.$$

In the next step define for $x = (x_1, x_2) \in \mathbb{R} \times [-w_\infty/2, w_\infty/2]$

$$F_n(x) := \begin{cases} \phi(x_2)\tilde{f}_n(x) + (1 - \phi(x_2))x, & \text{if } x_2 \leq 0 \\ \phi(x_2) f_n(x) + (1 - \phi(x_2))x, & \text{if } x_2 > 0. \end{cases}$$

The maps F_n induce maps $A_\infty \to X_n$ which fit together smoothly with $\psi_n|_{Y_\infty \backslash A_\infty}$ and $\tilde{\psi}_n|_{\tilde{Y}_\infty \backslash A_\infty}$, and in this way we obtain maps

$$\Psi_n: X_\infty \to X_n \setminus \left(\bigcup_{v \in I} \gamma_n^v \cup \bigcup_{v \in \tilde{I}} \tilde{\gamma}_n^v \right)$$

having the desired properties.   □

Proof of Proposition 5.1. Choose a sequence $(\delta_\mu)_{\mu \geq 1}$ of positive numbers converging to zero. For each $\mu \geq 1$ define a sequence of maps $(Y^i \to Y_n^i)_{n \geq 1}$, $i = 1, \ldots, N$, as in Claim 1 and 2 with $\delta = \delta_\mu$ and $r = \mu$. Using Claim 3, they can be joined together to form smooth maps

$$\psi_n^\mu: S \setminus \bigcup_{i=1}^{k'} \gamma^i \to S_n \setminus \bigcup_{i=1}^{k'} \gamma_n^i.$$

Then there is a diagonal sequence (ψ_n) with the properties claimed in part (iii). Note that the maps ψ_n are orientation preserving by construction and thus the convergence of the pulled back complex structures on S^* is clear from (iii).

In order to prove part (iv) of the proposition and to show how to get the maps φ_n we proceed as follows. We modify slightly the maps ψ_n near those degenerate boundary components of S^*, which do *not* result from collapsing non-degenerate boundary components in the sequence (S_n^*), such that the maps can be extended to these boundary components.

For that purpose we go back to the considerations in the proof of Claim 2 and show how to define the ψ_n near these boundary components.

We may assume that the pairs of pants Y_n, $n < \infty$, are degenerate and have the same signature. Their degenerate boundary components are γ_n^μ, $\mu \in I'$, where $I' \subset I \subset \{1, 2, 3\}$ is some non-empty subset of I. Now fix $\delta_0 := (1/2)\operatorname{arcsinh}(1)$. For any given $\delta < \delta_0$ construct ψ_n as in Claim 2. Then for sufficiently large n modify the restriction

$$\hat{\psi}_n: Y_\infty \setminus \bigcup_{\mu \in I'} \mathscr{C}(\gamma_\infty^\mu, \delta_0) \to Y_n$$

of ψ_n in the following way. Perturb $\widehat{\psi}_n$ near $\partial\mathscr{C}(\gamma_\infty^\mu, \delta_0)$ as in the proof of Claim 3 such that it fits together smoothly with the standard map

$$\beta_n^\mu \colon \mathscr{C}(\gamma_\infty^\mu, \delta_0) \cup \gamma_\infty^\mu \to \mathscr{C}(\gamma_n^\mu, \delta_0) \cup \gamma_n^\mu$$

and such that the resulting maps

$$Y_\infty \to Y_n \setminus \bigcup_{v \in I} \gamma_n^v \, ,$$

again denote by ψ_n, still satisfy Claim 2. The map β_n^μ is well-defined and unique by requiring that it maps $\mathscr{C}(\gamma_\infty^\mu, \delta_0)$ isometrically to $\mathscr{C}(\gamma_n^\mu, \delta_0)$ and its extension to the closure of the collars maps corners of $G_\infty[\delta_0]$ to corresponding corners of $G_n[\delta_0]$. Moreover, the maps β_n^μ are biholomorphic.

Hence we obtain diffeomorphisms

$$\psi_n \colon S \setminus \bigcup_{i=1}^{k'} \gamma^i \to S_n \setminus \bigcup_{i=1}^{k'} \gamma_n^i$$

having the properties claimed in part (iv). The maps φ_n are now defined by part (i) and by $\varphi_n|_{S_n \setminus \bigcup_{i=1}^{k'} \gamma_n^i} = \psi_n^{-1}$; and they are continuous. $\square$

For technical reasons we make the following

Remark 5.2. The last step in the proof shows that for φ_n in Proposition 5.1 we can make the additional requirement that there is an open set $U \subset S$ and for each $n \geq 1$ an open neighbourhood U_n of the degenerate boundary components of S_n^* in S_n such that $\varphi_n|_{U_n} \colon U_n \to U$ is j_n-j-biholomorphic.

Chapter V

The compactness theorem

In this chapter we formulate the compactness theorem for J-holomorphic curves and present Gromov's proof in detail. But first, we recall the example from the introduction.

Example 0.1. Let $S^2 = \mathbb{C} \cup \{\infty\}$ denote the Riemann sphere. Then $f_n: S^2 \to S^2 \times S^2$ with $f_n(z) := (z, 1/(n^2 z))$, $n \geq 1$, is a sequence of holomorphic curves in $S^2 \times S^2$. Each f_n is an embedding. As $n \to \infty$ the images $f_n(S^2)$ "converge" to $\bar{S} := S^2 \times \{0\} \cup \{0\} \times S^2$. In this way the simple closed loops $f_n(\{ |z| = 1/n \}) \subset S^2 \times S^2$ collapse to the point $(0, 0)$ at which the two spheres $S^2 \times \{0\}$ and $\{0\} \times S^2$ are glued together.

In the following section we first extend the notion of a closed J-holomorphic curve to encompass such degenerate curves as $\bar{S} \hookrightarrow S^2 \times S^2$, so-called cusp curves.

1. Cusp curves

In order to visualize the following construction, refer to Figure 18. Let S be a closed surface and let $(\gamma^i)_{i \in I}$ be a possibly empty family of finitely many smooth simple closed and pairwise disjoint loops in S. We mean by a smooth simple closed loop an embedded, 1-dimensional submanifold of S diffeomorphic to S^1. Let $\hat{S}$ be the surface obtained from $S \setminus \bigcup_{i \in I} \gamma^i$ by the one-point compactification at each end. By s'_k and s''_k, $k \in I$, we denote the two points of $\hat{S}$ added to $S \setminus \bigcup_{i \in I} \gamma^i$ at the two ends which arise from removing γ^k. We now identify s'_k with s''_k for $k \in I$. In this way we obtain a compact topological space $\bar{S}$. Hence we can get $\bar{S}$ from S by collapsing each loop γ^i to a point. By $\alpha: \hat{S} \to \bar{S}$ we denote the canonical projection. The points $\bar{s}_k := \alpha(s'_k)$ are called *singular points* of the *singular surface* $\bar{S}$. Via α the subset $\bar{S} \setminus \{ \bar{s}_i \mid i \in I \} \subset \bar{S}$ inherits a differentiable structure.

In the following, if S is a closed surface then $\bar{S}$ always denotes a singular surface obtained from S by the above construction and $\hat{S}$ the corresponding smooth surface from that construction. We call the components of $\hat{S}$ also the *the components of* $\bar{S}$. So $\hat{S}$ is the disjoint union of the components of $\bar{S}$. By $\mathrm{si}(\bar{S})$ we mean the set of singular points of $\bar{S}$ which is at most finite.

Definition. A *deformation* of S onto a singular surface $\bar{S}$ is a continuous surjective map $\varphi: S \to \bar{S}$ with the following properties:

(i) The preimage $\varphi^{-1}(\{\bar{s}\})$ of each singular point $\bar{s} \in \bar{S}$ is a simple closed loop.

(ii) $\varphi|_{S \setminus \varphi^{-1}(\mathrm{si}(\bar{S}))}$ is a diffeomorphism onto $\bar{S} \setminus \mathrm{si}(\bar{S})$. We denote its inverse map by φ^{-1}, that is $\varphi^{-1}: \bar{S} \setminus \mathrm{si}(\bar{S}) \to S \setminus \varphi^{-1}(\mathrm{si}(\bar{S}))$.

If $\varphi\colon S \to \bar{S}$ is a deformation and $U \subset \bar{S}$ a subset of $\bar{S}$ then $\varphi^{-1}(U)$ denotes the preimage of U under φ.

Definition. A complex structure $\bar{j}$ on a singular surface $\bar{S}$ is a complex structure on $\hat{S}$. We call the pair $(\bar{S}, \bar{j})$ a *singular Riemann surface* or *Riemann surface with singular points*. A continuous map $\bar{f}\colon (\bar{S}, \bar{j}) \to (M, J)$ to an almost complex manifold (M, J) is called J-holomorphic if $\bar{f} \circ \alpha$ is J-holomorphic. In this case $\bar{f}$ is called a *cusp curve* in M. The *area of a cusp curve* $\bar{f}$ is $\mathcal{A}(\bar{f}) := \mathcal{A}(\bar{f} \circ \alpha)$.

Definition. Let (M, J, μ) be an almost complex manifold with Hermitian metric, S a closed surface and $(j_n)_{n \geq 1}$ a sequence of complex structures on S. Assume that $f_n\colon (S, j_n) \to M$, $n \geq 1$, is a sequence of j_n-J-holomorphic curves in M. Then $(f_n)_{n \geq 1}$ is said to *converge weakly* to a cusp curve $\bar{f}\colon (\bar{S}, \bar{j}) \to (M, J)$ and $\bar{f}$ is called a *weak limit* of $(f_n)_{n \geq 1}$ if the following holds:

(i) For each $n \geq 1$ there is a deformation $\varphi_n\colon S \to \bar{S}$ such that $(\varphi_n^{-1})^* j_n$ converges in the C^∞-topology to $\bar{j}|_{\bar{S} \setminus \mathrm{si}(\bar{S})}$.

(ii) $(f_n \circ \varphi_n^{-1})_{n \geq 1}$ converges in C^∞ to $\bar{f}|_{\bar{S} \setminus \mathrm{si}(\bar{S})}$.

(iii) $\lim\limits_{n \to \infty} \mathcal{A}(f_n) = \mathcal{A}(\bar{f})$.

(iv) $(f_n \circ \varphi_n^{-1})_{n \geq 1}$ converges uniformly to $\bar{f}|_{\bar{S} \setminus \mathrm{si}(\bar{S})}$.

Remarks. Let $f_n\colon (S, j_n) \to M$, $n \geq 1$, be a sequence of closed j_n-J-holomorphic curves in (M, J, μ) and $\bar{f}\colon (\bar{S}, \bar{j}) \to M$ some weak limit of $(f_n)_{n \geq 1}$.

1. By (iv) the homotopy class of f_n is equal to the one of $\bar{f}$ for each large n.

2. Assume $\chi_n\colon (S, j_n) \to (S, j_n)$, $n \geq 1$, are conformal transformations. Then $(f_n \circ \chi_n)$ also converges weakly to $\bar{f}$. Moreover, the parametrization of the weak limit $\bar{f}$ is not unique. However, there exists the following uniqueness result:

Proposition 1.1. *Assume* $\bar{f}\colon \bar{S} \to M$ *is a weak limit of a sequence* (f_n) *of closed J-holomorphic curves. Then* $\bar{f}(\bar{S})$ *is the set of accumulation points of the sequences* $\{ (f_n(w_n)) \mid (w_n) \text{ is a sequence in } S \}$. *In particular, if* $\tilde{f}\colon \bar{S} \to M$ *is another weak limit of* (f_n) *then the images of* $\bar{f}$ *and* $\tilde{f}$ *coincide, i.e.,* $\bar{f}(\bar{S}) = \tilde{f}(\bar{S})$.

Proof. Let $\bar{f}\colon \bar{S} \to M$ be a weak limit of a sequence $f_n\colon S \to M$, $n \geq 1$, of closed j_n-J-holomorphic curves. Choose deformations $\varphi_n\colon S \to \bar{S}$ such that $(f_n \circ \varphi_n^{-1})$ converges uniformly to $\bar{f}|_{\bar{S} \setminus \mathrm{si}(\bar{S})}$. Let $A \subset M$ be the set of points $p \in M$ for which there exists a sequence $(w_n)_{n \geq 1}$ in S such that p is an accumulation point of $(f_n(w_n))$. We have to show $\bar{f}(\bar{S}) = A$. Note that $\bar{f}(\bar{S} \setminus \mathrm{si}(\bar{S})) \subset A$ since

$$\bar{f}(s) = \lim_{n \to \infty} f_n(\varphi_n^{-1}(s))$$

for each $s \in \bar{S} \setminus \mathrm{si}(\bar{S})$. Thus $\bar{f}(\bar{S}) \subset A$ since $A \subset M$ is a closed subset.

As usual, we denote by d the distance in M. Then, for $\varepsilon > 0$ and all sufficiently large n we have $d(f_n \circ \varphi_n^{-1}(s), \bar{f}(\bar{S})) \leq \varepsilon$ for each $s \in \bar{S} \setminus \mathrm{si}(\bar{S})$ since $(f_n \circ \varphi_n^{-1})$ converges uniformly. Thus $d(f_n(w), \bar{f}(\bar{S})) \leq \varepsilon$ for each $w \in S \setminus \varphi_n^{-1}(\mathrm{si}(\bar{S}))$ and thus for each $w \in S$ by continuity. This proves that $A \subset \bar{f}(\bar{S})$. $\qquad\square$

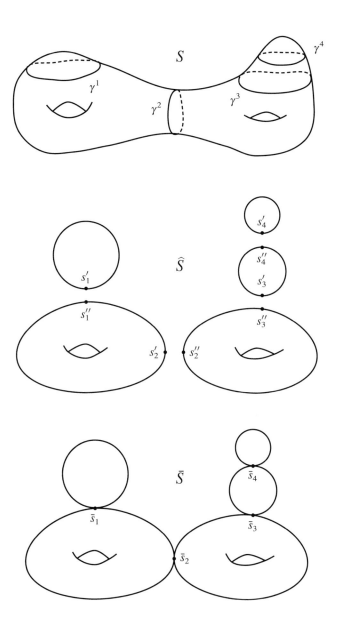

Figure 18.

We call a sequence (f_n) of J-holomorphic curves *of bounded area* if there is some constant $C < \infty$ with $\mathcal{A}(f_n) \leq C$ for each n. We can now state the compactness theorem for J-holomorphic curves.

Theorem 1.2 (Gromov's compactness theorem). *Let (M, J, μ) be a compact almost complex manifold with Hermitian metric, S a closed surface and (j_n) a sequence of complex structures on S. Assume $f_n \colon (S, j_n) \to M$ is a sequence of j_n-J-holomorphic curves of bounded area. Then there exists a subsequence of (f_n) converging weakly to some cusp curve $\bar{f} \colon \bar{S} \to M$.*

Remarks. **1.** Assume there exists a symplectic structure ω on M taming J. Then any sequence (f_n) of closed J-holomorphic curves in M representing the same homology class in $H_2(M, \mathbb{R})$ has bounded area by (II.2.1).

2. The collapsing of circles in minimal surfaces was discovered by J. Sacks and K. Uhlenbeck (see [SaU]).

2. Proof of the compactness theorem

Let (M, J, μ) be a compact almost complex manifold with Hermitian metric μ. We choose $0 < \varepsilon_0 < \operatorname{injrad}(M)$ as in Remark II.4.3. Hence (M, J, μ) together with ε_0 satisfies the assumptions of the Gromov-Schwarz lemma II.1.2 and the monotonicity lemma. We shall consider closed J-holomorphic curves in M whose areas are bounded by some fixed constant $C < \infty$ from above.

Convention 2.1. In this section *universal constant* always means a constant depending only on (M, J, μ), ε_0 and C; *universally bounded* means bounded by some universal constant.

Lemma 2.2. *For any closed J-holomorphic curve $f \colon S \to M$ with $\mathcal{A}(f) \leq C$ there exist finitely many points $s_1, \ldots, s_k$, whose number is universally bounded and some universal constant $\rho_0 > 0$ such that the following holds. With respect to the Poincaré metric h on $S^* := S \setminus \{s_1, \ldots, s_k\}$ the image under f of each ρ_0-ball in (S^*, h) is contained in some ε_0-ball in M.*

Note that S^* as in the lemma is certainly of non-exceptional type for $k > 2$ and recall Proposition IV.3.9 which states that (S^*, h) is a hyperbolic surface of finite area satisfying the assumption of Proposition IV.4.2 on the thick-thin decomposition.

Proof. Let $f \colon S \to M$ be a non-constant closed J-holomorphic curve with $\mathcal{A}(f) \leq C$. For $s \in S$ let $U(s)$ be the connected component of $f^{-1}\big(B_{\varepsilon_0/18}(f(s))\big)$ containing s. We can estimate $\mathcal{A}(f|_{U(s)})$ with the monotonicity lemma. Namely, there is no non-constant closed J-holomorphic curve in M whose image is contained in some ε_0-ball in M by Remark II.4.3. Hence the monotonicity lemma yields that

$$\mathcal{A}(f|_{U(s)}) \geq C_{\mathrm{ML}} \left(\frac{\varepsilon_0}{18} \right)^2.$$

Here C_{ML} is the constant from the monotonicity lemma. If $F \subset S$ is a subset such that the sets $U(s)$, $s \in F$, are pairwise disjoint then the number of points in F is bounded by $(18/\varepsilon_0)^2(\mathcal{A}(f)/C_{\mathrm{ML}})$ and hence universally bounded. Now choose such an F with a maximal number of elements. Thus F has the following property. For any $s \in S$ there is some $s' \in F$ with $U(s) \cap U(s') \neq \varnothing$. Since the image of f is not contained in any ε_0-ball in M, the number of points in F is greater than 2. Thus the Riemann surface $S^* := S \setminus F$ is of non-exceptional type and we denote by h its Poincaré metric.

If $A \subset S^*$ is an embedded closed annulus in S^* we claim that $d(f(s), f(\partial A)) < \varepsilon_0/6$ for each $s \in A$. Namely, we may assume that $U(s)$ is contained in the interior $A \setminus \partial A$ of A, for otherwise even $d(f(s), f(\partial A)) < \varepsilon_0/18 < \varepsilon_0/6$. By maximality of F for any $s \in A \setminus \partial A$ there is some $s_0 \in F$ with $U(s_0) \cap U(s) \neq \varnothing$. Since $A \subset S^* = S \setminus F$, the point s_0 is not contained in A and thus there is some $s' \in U(s_0) \cap \partial A$ and we obtain that

$$
\begin{aligned}
d(f(s), f(\partial A)) &\leq d(f(s), f(s')) \\
&\leq d(f(s), f(s_0)) + d(f(s_0), f(s')) \\
&< \frac{\varepsilon_0}{9} + \frac{\varepsilon_0}{18} = \frac{\varepsilon_0}{6} .
\end{aligned}
$$

Let $\partial_0 A$ and $\partial_1 A$ denote the components of ∂A. Since A is connected,

$$
\left\{ s \in A \mid d(f(s), f(\partial_0 A)) < \tfrac{\varepsilon_0}{6} \right\} \cap \left\{ s \in A \mid d(f(s), f(\partial_1 A)) < \tfrac{\varepsilon_0}{6} \right\} \neq \varnothing
$$

and thus $d(f(\partial_0 A), f(\partial_1 A)) < \varepsilon_0/3$. With the help of the following lemma, the proof of which we postpone, the proof of Lemma 2.2 can be concluded.

Lemma 2.3. *For any closed J-holomorphic curve $f\colon S \to M$ with $\mathcal{A}(f) \leq C$ there is a universal constant $\rho_0 > 0$ such that each ρ_0-ball in (S^*, h) is contained in some embedded annulus $A \subset S^*$ with $\ell(f|_{\partial_i A}) \leq \varepsilon_0/6$, $i = 0, 1$.*

For any given ρ_0-ball $B_{\rho_0}(s^*) \subset (S^*, h)$ now choose such a closed annulus $A \subset S^*$ as in the lemma. Then for any two points $s, s' \in B_{\rho_0}(s^*)$ we have that

$$
\begin{aligned}
d(f(s), f(s')) &\leq d(f(s), f(\partial A)) + \ell(f|_{\partial_0 A}) + \ell(f|_{\partial_1 A}) \\
&\quad + d(f(\partial_0 A), f(\partial_1 A)) + d(f(\partial A), f(s')) .
\end{aligned}
$$

Thus $d(f(s), f(s')) < \varepsilon_0$ and hence the diameter of $f(B_{\rho_0}(s^*))$ is smaller than ε_0. □

Proof of Lemma 2.3. Let $\phi\colon S^1 \times (0, T) \to M$ be a J-holomorphic map from the annulus $S^1 \times (0, T)$ with modulus $T < \infty$ to the compact almost complex manifold M. Since ϕ is conformal, and using the standard Euclidean metric on $S^1 \times (0, T)$, we can estimate

$$
\begin{aligned}
\mathcal{A}(\phi) &= \int_0^T \int_0^{2\pi} \|T\phi\|^2 \, d\theta \, dt \\
&\geq \frac{1}{2\pi} \int_0^T \left(\int_0^{2\pi} \|T\phi\| \, d\theta \right)^2 dt = \frac{1}{2\pi} \int_0^T \ell^2(\phi|_{S^1 \times \{t\}}) \, dt .
\end{aligned}
$$

Consequently, there is some $t_0 \in (0, T)$ with

$$\ell^2(\phi|_{S^1 \times \{t_0\}}) \leq \frac{2\pi \mathcal{A}(\phi)}{T}. \tag{2.1}$$

We shall see below that for a sufficiently small universal constant $\rho_0 > 0$ and any ρ_0-ball $B_{\rho_0}(s) \subset (S^*, h)$ there exist embedded annuli $A_0, A', A_1 \subset S^*$ with moduli Mod A_0, Mod A', Mod A_1, respectively, having the following properties:

(i) $B_{\rho_0}(s)$ is contained in A',

(ii) A' is closed and A_0, A_1 are semi-open,

(iii) Mod $A_0 < \infty$ and there exist conformal transformations

$$\phi_0: S^1 \times (0, \text{Mod } A_0] \to A_0$$
$$\phi': S^1 \times [\text{Mod } A_0, \text{Mod } A_0 + \text{Mod } A'] \to A'$$
$$\phi_1: S^1 \times [\text{Mod } A_0 + \text{Mod } A', \text{Mod } A_0 + \text{Mod } A' + \text{Mod } A_1) \to A_1$$

which fit together to a conformal embedding

$$\phi: S^1 \times (0, \text{Mod } A_0 + \text{Mod } A' + \text{Mod } A_1) \to S^*.$$

(iv) The moduli of A_0 and A_1 satisfy

$$\text{Mod } A_i \geq 36 \cdot \frac{2\pi C}{\varepsilon_0^2} \qquad \text{for } i = 0, 1.$$

The existence of such annuli concludes the proof. Namely, we may assume without loss of generality that Mod $A_1 < \infty$. By (2.1) there exist $t_0 \in [0, \text{Mod } A_0)$ and $t_1 \in [\text{Mod } A' + \text{Mod } A_0, \text{Mod } A' + \text{Mod } A_0 + \text{Mod } A_1)$ with

$$\ell^2(f \circ \phi_i|_{S^1 \times \{t_i\}}) \leq \frac{2\pi \mathcal{A}(f)}{\text{Mod } A_i} \leq \frac{\varepsilon_0^2}{36}.$$

Then put $A := \phi(S^1 \times [t_0, t_1])$. Hence $B_{\rho_0}(s) \subset A' \subset A$ and A has the desired properties. The existence of a ρ_0 with the above properties follows from the thick-thin decomposition of hyperbolic surfaces. Indeed, choose $\rho_0 > 0$ so small that $3\rho_0 < \sqrt{\rho_0}$ and

$$\log\left(\tanh \frac{\sqrt{\rho_0}}{2}\right) - \log\left(\tanh \frac{3\rho_0}{2}\right) \geq 36 \cdot \frac{2\pi C}{\varepsilon_0^2}. \tag{2.2}$$

Moreover, ρ_0 should be chosen so small that the following holds. If the injectivity radius injrad(S^*, s) in $s \in (S^*, h)$ is smaller than $2\sqrt{\rho_0}$ then

$$\text{injrad}(S^*, \cdot)|_{B_{\rho_0}(s)} < r < \text{arcsinh}(1) \tag{2.3}$$

for some r which satisfies

$$\frac{1}{\sinh r} - 1 \geq 36 \cdot \frac{2C}{\varepsilon_0^2}. \tag{2.4}$$

We claim that such a ρ_0 can be chosen depending only on ε_0 and C. The only property for which this is not obvious is (2.3). By Proposition IV.3.9 the surface (S^*, h) has finite area. The components of $\{\, s \in S^* \mid \text{injrad}(S^*, s) < \text{arcsinh}(1)\,\}$ are classified up to isometry in Proposition IV.4.2. From that classification it follows that ρ_0 satisfying (2.3) can be chosen depending only on ε_0 and C.

In order to construct the annuli A_0, A', A_1 for a given ρ_0-ball $B_{\rho_0}(s) \subset S^*$ as required, we distinguish whether s is contained in a "thick" or "thin" part of (S^*, h).

CASE 1. The injectivity radius in s satisfies $\text{injrad}(S^*, s) \geq 2\sqrt{\rho_0}$. Then choose $s' \in S^*$ with $d(s, s') = 2\rho_0$ and define

$$A_0 := B_{\sqrt{\rho_0}}(s') \setminus \overline{B}_{3\rho_0}(s')$$
$$A' := \overline{B}_{3\rho_0}(s') \setminus B_{\rho_0}(s')$$
$$A_1 := \overline{B}_{\rho_0}(s') \setminus \{s'\}.$$

Observe that ρ_0 was chosen such that $3\rho_0 < \sqrt{\rho_0}$ and thus $A_0, A', A_1 \subset B_{2\sqrt{\rho_0}}(s)$.

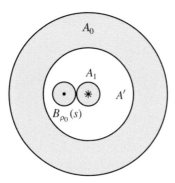

Figure 19.

Since $\text{injrad}(S^*, s) \geq 2\sqrt{\rho_0}$, the subsets A_0, A' and A_1 are indeed annuli and they obviously satisfy conditions (i), (ii) and (iii) from above. The ball $B_{2\sqrt{\rho_0}}(s)$ is isometric to a $2\sqrt{\rho_0}$-ball in the Riemannian covering space $\mathbb{H}$ of S^*. The calculations in Example I.5.7 yield

$$\text{Mod } A_0 = \log\left(\tanh \frac{\sqrt{\rho_0}}{2}\right) - \log\left(\tanh \frac{3\rho_0}{2}\right) \quad \text{and} \quad \text{Mod } A_1 = \infty.$$

Condition (2.2) for ρ_0 implies that

$$\text{Mod } A_0 \geq 36 \cdot \frac{2\pi C}{\varepsilon_0^2}$$

and consequently the annuli A_0, A_1 also satisfy condition (iv).

CASE 2. The injectivity radius in s satisfies $\text{injrad}(S^*, s) < 2\sqrt{\rho_0}$. Because of (2.3) and Proposition IV.4.2 about the thick-thin decomposition, some neighbourhood U of the ρ_0-ball $B_{\rho_0}(s)$ in S^* can be chosen such that

(a) either there is an $l < 2r$ such that U is isometric to

$$\left\{ w \in \langle T_l \rangle \backslash \mathbb{H} \mid \text{injrad}(\langle T_l \rangle \backslash \mathbb{H}, w) < \text{arcsinh}(1) \right\},$$

(b) or U is isometric to

$$\left\{ w \in \langle P \rangle \backslash \mathbb{H} \mid \text{injrad}(\langle P \rangle \backslash \mathbb{H}, w) < \text{arcsinh}(1) \right\}.$$

Now identify U with the corresponding subset of the hyperbolic cylinder $\langle T_l \rangle \backslash \mathbb{H}$ and parabolic cylinder $\langle P \rangle \backslash \mathbb{H}$, respectively.

If U is as in (a), then let A_0, A_1 be the two components of the subset

$$\left\{ w \in \langle T_l \rangle \backslash \mathbb{H} \mid r \leq \text{injrad}(\langle T_l \rangle \backslash \mathbb{H}, w) < \text{arcsinh}(1) \right\} \subset U$$

and A' the piece in between, that is

$$A' = \left\{ w \in \langle T_l \rangle \backslash \mathbb{H} \mid \text{injrad}(\langle T_l \rangle \backslash \mathbb{H}, w) \leq r \right\} \subset U.$$

Because of (2.3) we have that $B_{\rho_0}(s) \subset A'$.

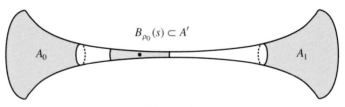

Figure 20.

The moduli of A_0 and A_1 were computed and estimated in Example I.5.5,

$$\text{Mod } A_0 = \text{Mod } A_1 \geq \pi \left(\frac{1}{\sinh r} - \frac{1}{\sinh \text{arcsinh}(1)} \right).$$

Thus, together with (2.4) it follows that

$$\text{Mod } A_0 = \text{Mod } A_1 \geq 36 \cdot \frac{2\pi C}{\varepsilon_0^2}.$$

The annuli $A_0, A', A_1 \subset U \subset S^*$ have the desired properties. If U is as in (b) then put

$$A_0 := \left\{ w \in \langle P \rangle \backslash \mathbb{H} \mid r \leq \mathrm{injrad}(\langle P \rangle \backslash \mathbb{H}, w) < \mathrm{arcsinh}(1) \right\}$$

$$A' := \left\{ w \in \langle P \rangle \backslash \mathbb{H} \mid r' \leq \mathrm{injrad}(\langle P \rangle \backslash \mathbb{H}, w) \leq r \right\}$$

$$A_1 := \left\{ w \in \langle P \rangle \backslash \mathbb{H} \mid \mathrm{injrad}(\langle P \rangle \backslash \mathbb{H}, w) \leq r' \right\},$$

where we choose $0 < r' < r$ such that $B_{\rho_0}(s) \subset A'$.

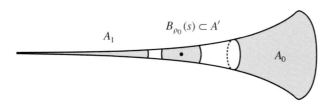

Figure 21.

Then $\mathrm{Mod}\, A_1 = \infty$ and

$$\mathrm{Mod}\, A_0 = \pi \left(\frac{1}{\sinh r} - \frac{1}{\sinh \mathrm{arcsinh}(1)} \right)$$

holds as we can see from Example I.5.6. The annuli A_0, A' and A_1 are contained in U and have the desired properties. ☐

Corollary 2.4. *Let $f: S \to (M, J, \mu)$ and (S^*, h) be as in Lemma 2.2. Then the differential Tf of f on S^* is universally bounded with respect to the Poincaré metric h. In particular $f|_{S^*}$ has a universal Lipschitz constant with respect to h.*

Proof. Let $\mathbb{H} \xrightarrow{p} S^*$ denote the Riemannian universal covering of S^*. The image of any ρ_0-ball $B_{\rho_0}(z) \subset \mathbb{H}$ under $f \circ p$ is contained in some ε_0-ball in M by Lemma 2.2. Hence the differential of $f \circ p$ is universally bounded with respect to the Poincaré metric λ of $B_{\rho_0}(z)$ by the Gromov-Schwarz lemma II.1.2. Thus it is universally bounded in z with respect to the metric of $\mathbb{H}$. But z was arbitrary. Therefore the differential of $f \circ p$ and thus of f is universally bounded. ☐

We now begin with the actual proof of the compactness theorem. Let $f_n \colon (S, j_n) \to M$, $n \geq 1$, be a sequence of closed j_n-J-holomorphic curves in the compact manifold (M, J, μ) with $\mathcal{A}(f_n) \leq C$ for each n, where S is a closed surface of genus g and (j_n) a sequence of complex structures on S. By S_n we denote the Riemann surface (S, j_n).

Choice of subsequence. Using Lemma 2.2 and Corollary 2.4, choose for each $n \geq 1$ a subset $F_n \subset S_n$ such that $f_n|_{S_n^*}$ has a Lipschitz constant independent of n, with respect to the Poincaré metric h_n on $S_n^* := S_n \setminus F_n$, and that the number k_n of points in F_n is universally bounded.

After passing to some subsequence of (f_n), again denoted by (f_n), we may assume that $k_n =: k$ is independent of n. Thus the S_n^* are hyperbolic surfaces of the same signature $(g, 0, k)$. By Proposition IV.4.2, the simple closed geodesics γ_n^i, $i = 1, \ldots, l_n$, in (S_n^*, h_n), whose length is smaller than $2 \operatorname{arcsinh}(1)$, are pairwise disjoint. Their number l_n is bounded by $3g - 3 + k$. After passing to a further subsequence we can suppose that $l_n =: l$ is independent of n and that for each $n > 1$ there exists an orientation preserving diffeomorphism $\chi_n \colon S_1^* \to S_n^*$ with $\chi_n(\gamma_1^i) = \gamma_n^i$ for $i = 1, \ldots, l$. This is possible since there are only finitely many pairs of pants decompositions (up to pairs of pants preserving diffeomorphisms) of a surface of signature $(g, 0, k)$. Moreover, it can be assumed that the sequences $(\ell(\gamma_n^i))_{n \geq 1}$, $i = 1, \ldots, l$, are convergent. Let $I \subset \{1, \ldots, l\}$ denote the subset of indices i with $\ell(\gamma_n^i) \to 0$ as $n \to \infty$. Let $S_n^1, \ldots, S_n^{m_0} \subset S_n^*$ be the closures of the components of $S_n^* \setminus \bigcup_{i \in I} \gamma_n^i$ in S_n^*. We may assume that $\chi_n(S_1^\nu) = S_n^\nu$ for $\nu = 1, \ldots, m_0$. As in Section 1, collapsing the loops γ_1^i in S for $i \in I$, a singular surface $\bar{S}$ is obtained. The next claim is more or less a restatement of Proposition IV.5.1.

Claim. There is a subsequence of (S_n), again denoted by $(S_n)_{n \geq 1}$, a complex structure $\bar{\jmath}$ on $\bar{S}$, a finite subset $F \subset \bar{S} \setminus \operatorname{si}(\bar{S})$ and for each $n \geq 1$ a deformation $\varphi_n \colon S_n \to \bar{S}$ such that the following holds:

(i) φ_n maps F_n bijectively onto F,

(ii) $(\varphi_n^{-1})^* h_n$ converges in C^∞ to the Poincaré metric h of $\bar{S} \setminus (F \cup \operatorname{si}(\bar{S}))$ as $n \to \infty$,

(iii) $(\varphi_n^{-1})^* j_n$ converges in C^∞ to $\bar{\jmath}|_{\bar{S} \setminus \operatorname{si}(\bar{S})}$ as $n \to \infty$.

Namely, the subsequence (S_n), the corresponding deformations φ_n and the singular Riemann surface are obtained in the following way. For $\nu = 1, \ldots, m_0$ and the originally given sequences $(S_n^1)_{n \geq 1}, \ldots, (S_n^{m_0})_{n \geq 1}$ apply Proposition IV.5.1, passing successively to subsequences.

So let $f_n \colon S_n \to M$, $n \geq 1$, be the corresponding subsequence. By Corollary 2.4 and since $(\varphi_n^{-1})^* h_n$ converges in C^∞ to the Poincaré metric h of $\bar{S} \setminus (F \cup \operatorname{si}(\bar{S}))$ the maps $f_n \circ \varphi_n^{-1}|_{\bar{S} \setminus (F \cup \operatorname{si}(\bar{S}))}$, $n \geq 1$, are uniformly (with respect to n) Lipschitz on each compact subset in the domain. Since M is compact, some subsequence of $(f_n \circ \varphi_n^{-1})$ restricted to $\bar{S} \setminus (F \cup \operatorname{si}(\bar{S}))$ converges in C^0 to a continuous map

$$\bar{f} \colon \bar{S} \setminus (F \cup \operatorname{si}(\bar{S})) \to M$$

by Arzelà-Ascoli's theorem (see [Lg1]). Again, denote this subsequence by $(f_n \circ \varphi_n)$. Since $(\varphi_n^{-1})^* j_n$ converges in C^∞ on $\bar{S} \setminus (F \cup \operatorname{si}(\bar{S}))$, Corollary III.3.2 implies that the $(\varphi_n^{-1})^* j_n$-J-holomorphic maps $f_n \circ \varphi_n^{-1}$ restricted to $\bar{S} \setminus (F \cup \operatorname{si}(\bar{S}))$ converge in C^∞ to $\bar{f}$ as $n \to \infty$ and thus $\bar{f}$ is $\bar{\jmath}$-J-holomorphic.

Convergence of area. From Proposition IV.2.8 and the construction of $\bar{S}$ we see that

$$\int_{\bar{S} \setminus (F \cup \operatorname{si}(\bar{S}))} \sigma_h = \int_{\bar{S} \setminus (F \cup \operatorname{si}(\bar{S}))} \sigma_{(\varphi_n^{-1})^* h_n} = 2\pi(2g - 2 + k) < \infty$$

for each n. Here σ_h and $\sigma_{(\varphi_n^{-1})^* h_n}$ are the volume forms induced by the metrics h and $(\varphi_n^{-1})^* h_n$, respectively. Moreover, for each compact subset $K \subset \bar{S} \setminus (F \cup \mathrm{si}(\bar{S}))$

$$\lim_{n \to \infty} \int_K \sigma_{(\varphi_n^{-1})^* h_n} = \int_K \sigma_h ,$$

since $(\varphi_n^{-1})^* h_n$ converges uniformly on K. Summing up, this implies that for any $\varepsilon > 0$ there is an open neighbourhood U of $F \cup \mathrm{si}(\bar{S})$ in $\bar{S}$ such that

$$\int_{U \setminus (F \cup \mathrm{si}(\bar{S}))} \sigma_h \le \varepsilon \qquad \text{and} \qquad \int_{U \setminus (F \cup \mathrm{si}(\bar{S}))} \sigma_{(\varphi_n^{-1})^* h_n} \le \varepsilon$$

for each sufficiently large n. Since Tf_n is universally bounded on S_n^* with respect to h_n, there is some constant c such that for each sufficiently large n

$$\mathcal{A}(f_n \circ \varphi_n^{-1}|_{U \setminus \mathrm{si}(\bar{S})}) \le c \cdot \varepsilon . \tag{2.5}$$

The sequence $(f_n \circ \varphi_n^{-1})$ restricted to $\bar{S} \setminus U$ converges in C^1, hence

$$\lim_{n \to \infty} \mathcal{A}(f_n \circ \varphi_n^{-1}|_{\bar{S} \setminus U}) = \mathcal{A}(\bar{f}|_{\bar{S} \setminus U})$$

and together with (2.5) we obtain that

$$\limsup_{n \to \infty} \mathcal{A}(f_n \circ \varphi_n^{-1}) \le \mathcal{A}(\bar{f}) .$$

Convergence in C^1 implies that

$$\liminf_{n \to \infty} \mathcal{A}(f_n \circ \varphi_n^{-1}) \ge \mathcal{A}(\bar{f}) .$$

Combining the last two estimates, this proves the convergence of area,

$$\lim_{n \to \infty} \mathcal{A}(f_n \circ \varphi_n^{-1}) = \mathcal{A}(\bar{f}) .$$

Extension to $\bar{S}$. Up to now the $\bar{j}$-J-holomorphic map $\bar{f}$ is only defined on $\bar{S} \setminus (F \cup \mathrm{si}(\bar{S}))$. Since the area of $\bar{f}$ is finite, $\bar{f}$ can be extended to a J-holomorphic map defined on $\hat{S}$ by Theorem III.2.1 on the removal of singularities. Let $\alpha: \hat{S} \to \bar{S}$ be the canonical projection as in Section 1. To conclude the proof of the compactness theorem it remains to show that:

(a) The map $\bar{f}$ is compatible with the identification α, hence $\bar{f}$ can be viewed as $\bar{j}$-J-holomorphic map $\bar{f}: \bar{S} \to M$,

(b) $(f_n \circ \varphi_n^{-1})$ converges uniformly to $\bar{f}|_{\bar{S} \setminus \mathrm{si}(\bar{S})}$ and

(c) also converges in the C^∞-topology.

To prove (a) we argue indirectly. Consider $\bar{f}: \widehat{S} \to M$ and assume that there exists an $\bar{s} \in \text{si}(\bar{S})$ such that $\bar{f}(s') \neq \bar{f}(s'')$ for the preimages $s', s'' \in \widehat{S}$ of $\bar{s}$ under α. We define

$$\rho := \min\{\varepsilon_0, \tfrac{1}{4}d(\bar{f}(s'), \bar{f}(s''))\}\,.$$

Then choose compact, connected neighbourhoods U' and U'' of s' and s'' in $\widehat{S}$, respectively, with $\bar{f}(U') \subset B_\rho(\bar{f}(s'))$ and $\bar{f}(U'') \subset B_\rho(\bar{f}(s''))$. After possibly making U' and U'' smaller we can assume that

$$\mathscr{A}(f_n|_{\varphi_n^{-1} \circ \alpha(U' \cup U'')}) < C_{\text{ML}} \cdot \rho^2 \tag{2.6}$$

for each n. Since $(f_n \circ \varphi_n^{-1})$ restricted to $\bar{S} \setminus (F \cup \text{si}(\bar{S}))$ converges in C^0, we have

$$f_n\big(\partial(\varphi_n^{-1} \circ \alpha(U' \cup U''))\big) \subset B_\rho(\bar{f}(s')) \cup B_\rho(\bar{f}(s''))$$

for each sufficiently large n. Hence (2.6) contradicts the monotonicity lemma. This proves (a).

For the proof of (b) we also argue indirectly. It is already known that $(f_n \circ \varphi_n^{-1})$ restricted to $\bar{S} \setminus (F \cup \text{si}(\bar{S}))$ converges in C^0. So assume that $(f_n \circ \varphi_n^{-1})$ does not converge uniformly on $\bar{S} \setminus \text{si}(\bar{S})$. Then there exists some $s_0 \in F \cup \text{si}(\bar{S})$, some sequence $(s_\nu)_{\nu \geq 1}$ in $\bar{S} \setminus \text{si}(\bar{S})$ converging to s_0 in $\bar{S}$ and a subsequence $(f_{n_\nu} \circ \varphi_{n_\nu}^{-1})_{\nu \geq 1}$ of $(f_n \circ \varphi_n^{-1})_{n \geq 1}$ such that

$$d(f_{n_\nu} \circ \varphi_{n_\nu}^{-1}(s_\nu), \bar{f}(s_\nu)) > \varepsilon \tag{2.7}$$

for some $\varepsilon \in (0, \varepsilon_0)$ and all $\nu \geq 1$. Now pick some compact neighbourhood $U \subset \bar{S}$ of s_0 with $\partial U \cap (F \cup \text{si}(\bar{S})) = \varnothing$ such that

$$\bar{f}(U) \subset B_{\varepsilon/4}(\bar{f}(s_0)) \qquad \text{and}$$
$$\mathscr{A}(\bar{f}|_U) < C_{\text{ML}} \cdot (\tfrac{\varepsilon}{4})^2\,. \tag{2.8}$$

On the other hand by the continuity of $\bar{f}$, by the C^0-convergence of $(f_n \circ \varphi_n^{-1})$ outside of $F \cup \text{si}(\bar{S})$ and the choice of U we have for all sufficiently large ν

$$f_{n_\nu} \circ \varphi_{n_\nu}^{-1}(s_\nu) \notin B_{\varepsilon/2}(\bar{f}(s_0)) \qquad \text{and}$$
$$f_{n_\nu} \circ \varphi_{n_\nu}^{-1}(\partial U) \subset B_{\varepsilon/4}(\bar{f}(s_0))\,.$$

By the monotonicity lemma

$$\mathscr{A}(f_{n_\nu}|_{\varphi_{n_\nu}^{-1}(U)}) \geq C_{\text{ML}} \cdot (\tfrac{\varepsilon}{4})^2$$

in contradiction to (2.8), since $\mathscr{A}(f_{n_\nu}|_{\varphi_{n_\nu}^{-1}(U)}) \to \mathscr{A}(\bar{f}|_U)$ as $\nu \to \infty$. Hence $(f_n \circ \varphi_n^{-1})$ converges uniformly to $\bar{f}|_{\bar{S} \setminus \text{si}(\bar{S})}$.

To prove (c) the reader may apply Proposition III.4.1 to conclude that the uniformly convergent sequence $(f_n \circ \varphi_n^{-1})$ converges even in C^∞. Without using Proposition III.4.1, (c) can be proved in the following way. Recall that C^∞-convergence on $\bar{S} \setminus (F \cup \text{si}(\bar{S}))$ was already clear. By Remark IV.5.2 the φ_n may have been chosen so that additionally the φ_n^{-1} are $\bar{j}$-j_n-holomorphic in some neighbourhood U of F. Thus $(f_n \circ \varphi_n^{-1})$ converges in C^∞ also near F by Proposition III.3.1. This also proves (c) and concludes the proof of the compactness theorem.

Remarks. 1. Of course, if (M, J) is non-compact then the analogous assertion of the compactness theorem holds under the additional assumption that the J-holomorphic curves in consideration are contained in a fixed compact subset of M.

2. Assume (J_n) is a sequence of almost complex structures on a compact manifold M, converging in C^∞ to J. Then the compactness theorem generalizes to sequences (f_n) of closed j_n-J_n-holomorphic curves of bounded area, where the area is measured with respect to a fixed Hermitian metric on (M, J). Any limit cusp curve is J-holomorphic where the definition of weak convergence generalizes in the obvious way. Indeed, in this slightly more general situation the proof of the compactness theorem remains still valid because of the remarks in Section III.4.

3. Bubbles

In this section we shall have a closer look at weakly converging sequences of closed J-holomorphic curves $f_n: (S, j) \to (M, J, \mu)$, $n \geq 1$, in a compact manifold (M, J, μ) where the complex structure j on S does not depend on n.

Lemma 3.1. *Let (S, j) be a closed Riemann surface and $F \subset S$ a finite subset of S such that $S \setminus F$ is of non-exceptional type. Let h denote the Poincaré metric of $S \setminus F$. Then there is a constant $c > 0$ depending only on (S, j) such that any simple closed geodesic in $S \setminus F$ of length smaller than c is homotopically trivial.*

Proof. Each annulus $A \subset (S, j)$ represents a free homotopy class $[A]$ of a simple closed unparametrized loop in S. Since S is compact, there are only finitely many such classes and

$$\sup \left\{ \operatorname{Mod} A \mid A \subset S \text{ is an annulus, } [A] \text{ is non-trivial} \right\} < \infty.$$

Proposition IV.4.2 on thick-thin decomposition and Example I.5.5 show that each closed geodesic in $(S \setminus F, h)$ of length $l < 2 \operatorname{arcsinh}(1)$ is contained in an annulus $A \subset S \setminus F \subset S$ with modulus

$$\operatorname{Mod} A \geq \pi \left((\sinh \tfrac{l}{2})^{-1} - 1 \right).$$

This concludes the proof. □

Before proceeding, we recall some notation. If S is a closed surface then $\bar{S}$ always denotes a singular surface obtained from S by collapsing finitely many pairwise disjoint simple closed loops; $\widehat{S}$ is the disjoint union of the components of $\bar{S}$ and $\operatorname{si}(\bar{S})$ is the set of singular points of $\bar{S}$.

Any homotopically trivial simple closed loop γ in S bounds a disc by Remark IV.3.4 which is an immediate consequence of the Jordan-Schoenflies theorem (see [Mo]). Consequently, we obtain

Lemma 3.2. *Let f_n: $(S, j) \to M$, $n \geq 1$, be a weakly converging sequence of closed J-holomorphic curves. Then its weak limit can be parametrized by a singular Riemann surface $(\bar{S}, \bar{j})$ where $\widehat{S}$ is diffeomorphic to a disjoint union of S with at most finitely many 2-spheres.*

The next theorem will in fact imply the following

Addition to Lemma 3.2. *Under the above assumptions, $(\widehat{S}, \bar{j})$ is conformally equivalent to a disjoint union of (S, j) with at most finitely many 2-spheres.*

Theorem 3.3. *Let (E, J, μ) denote a compact almost complex manifold with Hermitian metric, (S, j) a closed Riemann surface and $E \xrightarrow{\pi} S$ a fibration of E over S which is J-j-holomorphic. Then any sequence of j-J-holomorphic sections of $E \xrightarrow{\pi} S$ with bounded area has some subsequence $(f_n)_{n \geq 1}$ with the following properties:*

(i) *There exists a finite subset $Z \subset S$ and a j-J-holomorphic section f_∞ of $E \xrightarrow{\pi} S$ such that $(f_n)_{n \geq 1}$ restricted to $S \setminus Z$ converges in C^∞ to $f_\infty|_{S \setminus Z}$ and*

(ii) *each fibre E_z, $z \in Z$, contains at least one non-constant J-holomorphic curve ϕ_z: $S^2 \to E_z$, that is a rational J-curve, passing through $f_\infty(z)$. The sequence (f_n) converges weakly to a cusp curve in E whose image is the union of $f_\infty(S)$, the images of ϕ_z, $z \in Z$, and at most finitely many images of further rational J-curves $S^2 \to E_z$, $z \in Z$.*

Proof. For any given sequence of j-J-holomorphic sections there exists a subsequence f_n: $(S, j) \to E$, $n \geq 1$, converging weakly to some cusp curve $\bar{f}$: $(\bar{S}, \bar{j}) \to E$. Let φ_n: $S \to \bar{S}$, $n \geq 1$, be a sequence of deformations, such that $(f_n \circ \varphi_n^{-1})$ converges uniformly to $\bar{f}|_{\bar{S} \setminus \text{si}(\bar{S})}$. By Lemma 3.2

$$\widehat{S} = \left(\bigsqcup_{i \in I} S_i^2 \right) \sqcup S$$

is a disjoint union of S with at most finitely many 2-spheres S_i^2, $i \in I$. As usual α: $\widehat{S} \to \bar{S}$ denotes the canonical projection. Since π: $E \to S$ is J-j-holomorphic,

$$\pi \circ \bar{f} \circ \alpha: (\widehat{S}, \bar{j}) \to (S, j)$$

is a holomorphic map. We let Σ_1 denote the union of the components of $\widehat{S}$ on which $\pi \circ \bar{f} \circ \alpha$ is constant and put $\Sigma_2 := \widehat{S} \setminus \Sigma_1$. The holomorphic map

$$\beta := \pi \circ \bar{f} \circ \alpha|_{\Sigma_2}: \Sigma_2 \to S$$

is surjective by Proposition 1.1 since each sequence $(f_n(s))_{n \geq 1}$, $s \in S$, has some accumulation point in the compact fibre E_s.

We claim that β is also injective. Arguing indirectly we assume there exist two different points $s_1, s_2 \in \Sigma_2$ with $\beta(s_1) = \beta(s_2)$. Since β as a non-constant holomorphic map is open, we may assume that $\alpha(s_1)$, $\alpha(s_2)$ are non-singular points. Then we can choose disjoint neighbourhoods U_1 and U_2 of $\alpha(s_1)$ and $\alpha(s_2)$ in $\bar{S} \setminus \text{si}(\bar{S})$, respectively, such that

$$\pi \circ \bar{f}(U_1) = \pi \circ \bar{f}(U_2). \tag{3.1}$$

On the other hand, since the f_n are sections of $E \overset{\pi}{\to} S$, we have that

$$\pi \circ f_n \circ \varphi_n^{-1}(U_v) = \varphi_n^{-1}(U_v)$$

for $v = 1, 2$ and thus

$$\pi \circ f_n \circ \varphi_n^{-1}(U_1) \cap \pi \circ f_n \circ \varphi_n^{-1}(U_2) = \varnothing.$$

Since $(f_n \circ \varphi_n^{-1})$ converges uniformly and $\pi \circ \bar{f}(U_1)$ is open, this contradicts (3.1). Hence $\beta : (\Sigma_2, \bar{j}|_{\Sigma_2}) \to (S, j)$ is biholomorphic.

In particular it has been shown that $\bar{f}(\alpha(\Sigma_2))$ is the image of a J-holomorphic section $f_\infty : S \to E$ of $E \overset{\pi}{\to} S$ since $\pi \circ \bar{f} \circ \alpha \circ \beta^{-1} = \text{id}_S$.

We put $Z := \pi \circ \bar{f}(\text{si}(\bar{S}))$. Since $(f_n \circ \varphi_n^{-1})$ converges uniformly to $\bar{f}$, the functions

$$S \setminus Z \ni s \mapsto d(f_n(s), \bar{f}(\bar{S}))$$

converge uniformly to zero as $n \to \infty$. For any compact subset $K \subset S \setminus Z$ we have $d(\pi^{-1}(K), \pi^{-1}(Z)) > 0$ and thus the functions

$$S \setminus Z \ni s \mapsto d(f_n(s), f_\infty(S))$$

converge in C^0 to zero as $n \to \infty$. Since f_∞ is also a section of $E \overset{\pi}{\to} S$, this implies that the maps $f_n|_{S \setminus Z}$ converge in C^0 to $f_\infty|_{S \setminus Z}$, and thus as j-J-holomorphic maps also in C^∞ by Proposition III.3.1.

To conclude the proof it remains to show that for any $z \in Z$ the set $\bar{f}(\bar{S}) \cap E_z$ contains some rational J-curve. For that purpose we may assume that $\bar{f}$ is constructed as in the proof of the compactness theorem in the last section.

For each n let $F_n \subset S$ be a maximal subset of S such that for $s \in F_n$ the connected subsets $f_n^{-1}\big(B_{\varepsilon_0/18}(f_n(s))\big) = \pi\big(B_{\varepsilon_0/18}(f_n(s))\big)$ of S are pairwise disjoint. Without loss of generality we may further assume that the number of points in F_n is the same for each n. By construction the preimages of $\text{si}(\bar{S})$ under φ_n are closed geodesics on $S \setminus F_n$ with respect to the Poincaré metric. Now pick any $\bar{s} \in \text{si}(\bar{S})$ with $\pi \circ \bar{f}(\bar{s}) = z$. From Lemma 3.1 it follows that for each sufficiently large n the closed geodesic $\varphi_n^{-1}(\{\bar{s}\}) \subset S \setminus F_n$ is homotopically trivial in S. Hence it bounds a closed disc Δ_n which is mapped under φ_n to the union of finitely many 2-spheres $S_1^2, \dots, S_k^2 \subset \bar{S}$ contained in $\alpha(\Sigma_1)$. Note that $\varphi_n(\Delta_n)$ is independent of n and that $\pi \circ \bar{f}$ is constant and equal to z on the connected set $\varphi_n(\Delta_n)$. Since $\partial\Delta_n$ is a geodesic in $S \setminus F_n$, the disc Δ_n contains at least two points of F_n. However, their images in E under f_n are at least $\varepsilon_0/18$ apart from each other by definition of F_n. Since $(f_n \circ \varphi_n^{-1})$ converges uniformly, $\bar{f}$ is non-constant on at least one of the spheres $S_1^2, \dots, S_k^2$. $\qquad\square$

Remarks. 1. To prove the Addition to 3.2 choose $E = (S \times M, j \oplus J)$ with the projection π onto the first factor. For a given weakly converging sequence $f_n: (S, j) \to M$, $n \geq 1$, of closed J-holomorphic curves consider the corresponding sequence of graphs (Γ_{f_n}) which are sections of $E \xrightarrow{\pi} S$. Let $\bar{\Gamma}$ be a weak limit of a subsequence of (Γ_{f_n}). Then (f_n) converges weakly to the composition of $\bar{\Gamma}$ with the projection onto M.

2. A subsequence (f_n) of a sequence of J-holomorphic sections of $E \xrightarrow{\pi} S$ of bounded area converges on $S \setminus Z$ in C^∞ to a J-holomorphic section f_∞ where Z is as in the theorem. The area of f_∞ satisfies

$$\mathscr{A}(f_\infty) \leq \limsup_{n \to \infty} \mathscr{A}(f_n)$$

and equality holds if and only if $Z = \varnothing$. When passing to the limit in each point $z \in Z$ some area is lost. This area is caught up in rational J-holomorphic curves $S^2 \to E_z$ called *bubbles* in z. Considering again the special case $E = (S \times M, j \oplus J)$, the analogous statements for sequences of closed J-holomorphic curves in compact manifolds is obtained.

3. Assume the compact manifold (M, J) does not contain any rational J-curve. Then any sequence of closed j-J-holomorphic curves $f_n: (S, j) \to (M, J)$ of bounded area has some uniformly convergent subsequence.

Recall that in order to find some weakly converging subsequence of a given sequence of closed J-holomorphic curves, certain points in each domain were removed. Finally, using the results proved so far, we want to illustrate the role these points play in the detection of bubbles.

Let $f_n: (S, j) \to (M, J, \mu)$ be a sequence of closed J-holomorphic curves weakly converging to a cusp curve $\bar{f}: \bar{S} \to M$. We assume that some bubble arises as $n \to \infty$ and that $\bar{f}$ is not a single bubble. The latter means that for instance the case $S^2 \ni z \to 1/nz \in S^2, n \geq 1$, is excluded.

In order to construct $\bar{f}$ we first removed a finite set F_n in the domain of f_n in the following way. For $s \in S$ a distinguished neighbourhood was defined, now we call it $U_n(s)$, namely the connected component of $f_n^{-1}(B_{\varepsilon_0/18}(f_n(s)))$ containing s. The finite set F_n was a maximal set such that the neighbourhoods $U_n(s), s \in F_n$, were pairwise disjoint.

Claim. There arises some bubble at $z \in S$ if and only if for any neighbourhood V of z each set $F_n \cap V$ contains more than one point, provided n is sufficiently large.

Indeed, let z be a point at which some bubble arises and pick any open neighbourhood V of z. Such a bubble has diameter greater than ε_0 by Remark II.4.3. From Theorem 3.3 applied to $E = S \times M$, we deduce that the diameter of $f_n(V)$ is greater than ε_0 for each sufficiently large n. By the choice of F_n one direction of the claim follows. Conversely, assume $z \in S$ is some point such that for any neighbourhood V of z the set $F_n \cap V$ contains more than one point for each sufficiently large n. Then (f_n) restricted to any neighbourhood of z does not converge uniformly since the image under f_n of any path in S connecting two different points in F_n is longer than $\varepsilon_0/18$. Hence there develops some bubble at z.

Claim. Let $z \in S$ be a point at which some bubble occurs. Then for any $r > 0$ and any neighbourhood V of z and each sufficiently large n there exists an annulus $A \subset V \cap (S \setminus F_n)$ of modulus greater than r which is not contained in any parabolic annulus $A' \subset S \setminus F_n$.

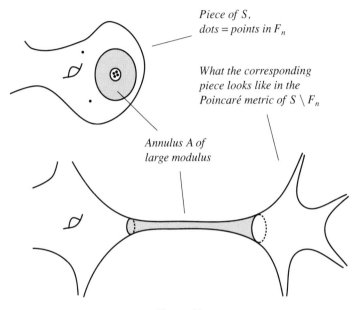

Piece of S,
dots = points in F_n

What the corresponding
piece looks like in the
Poincaré metric of $S \setminus F_n$

Annulus A of
large modulus

Figure 22.

To prove this we may assume that V is some small open disc and pick some conformal transformation $\psi \colon D \to V$ from the open unit disc in $\mathbb{C}$ to V mapping 0 to s_0. Choose $0 < \varepsilon < 1$ such that $-\log \varepsilon > r$. For any positive integer k let A_k be the annulus $A_k = \{ z \in D \mid \varepsilon^k < |z| < \varepsilon^{k-1} \}$. Then $\psi(A_k) \subset V$ has modulus equal to $-\log \varepsilon > r$ by Example I.5.4. Since the number of points in F_n is uniformly bounded by some constant $N < \infty$ for each n, there exists an annulus $A \in \{ \psi(A_k) \mid 1 \leq k \leq N+1 \}$ which does not contain any point of F_n. If n is sufficiently large the component of $S \setminus A$ contained in V contains more than one point of F_n by the last claim. Since V is some small disc and $\bar{f}$ is assumed not to be a single bubble, $S \setminus V$ contains more than one point of F_n. This proves that A is not contained in any parabolic annulus. In other words, the elements in the fundamental group of $S \setminus F_n$ associated to the free homotopy class $[A]$ are hyperbolic.

Claim. Let $z \in S$ be some point at which some bubble arises and V any neighbourhood of z in S. Then for any $\varepsilon > 0$ and each sufficiently large n there exists a closed geodesic γ in $S \setminus F_n$ with its Poincaré metric having the following properties. The closed geodesic γ has length smaller than ε and bounds a disc Δ in S satisfying $z \in \mathrm{int}\, \Delta \subset V$.

Namely, for r as large as we wish and each n sufficiently large we can choose an annulus A of modulus $> r$ as in the last claim. We write $S \setminus F_n = \Sigma \setminus \mathbb{H}$ as a quotient of the hyperbolic plane by a subgroup Σ of $\mathrm{Iso}^+(\mathbb{H})$ and let γ denote the unparametrized, simple closed geodesic in $S \setminus F_n$ representing $[A]$. Then $A = \langle \sigma \rangle \setminus \tilde{A}$ for some connected subset $\tilde{A}$ of $\mathbb{H}$ and some hyperbolic isometry $\sigma \in \Sigma$. By Remark I.5.3

$$\frac{\ell(\gamma)}{2\pi} = \frac{1}{\mathrm{Mod}\, \langle \sigma \rangle \setminus \mathbb{H}} \leq \frac{1}{\mathrm{Mod}\, A} = \frac{1}{r}.$$

Since $\bar{f}$ is assumed not to be a single bubble, one easily sees that the modulus of any hyperbolic annulus $A' \subset S \setminus (F_n \cup V)$ with $[A'] = [A]$ is bounded from above independently of n. Thus $\gamma \subset V$ and $\ell(\gamma) < \varepsilon$ provided r is large enough.

Summing up we see that there exists a sequence (γ_n) of simple closed geodesics in $S \setminus F_n$ which converges to a point $z \in S$ at which some bubble develops. Moreover, γ_n bounds some disc in S containing z in its interior provided n is sufficiently large. In the limit surface $\bar{S}$ these closed geodesics are collapsed to singular points.

Finally we shall give two simple examples of weakly converging sequences of J-holomorphic curves where more than one bubble arises at one point. They are simple modifications of the example at the beginning of this chapter. The reader can figure out how the points in F_n are distributed and detect the short simple closed geodesics in $S \setminus F_n$.

Example 3.4. Put $E = S^2 \times S^2 \times S^2$. With respect to the projection onto the first factor consider the sequence of sections (f_n) given by

$$f_n(z) = \left(z, \frac{1}{nz}, \frac{1}{n^2 z} \right).$$

Outside 0 the sequence converges in C^∞ to the section $f_\infty(z) = (z, 0, 0)$. At 0 two bubbles develop. The sequence (f_n) converges weakly to a cusp curve with image

$$S^2 \times \{0\} \times \{0\} \;\cup\; \{0\} \times S^2 \times \{0\} \;\cup\; \{0\} \times \{\infty\} \times S^2 \subset E.$$

Only one bubble, namely $\{0\} \times S^2 \times \{0\}$, is attached at $f_\infty(0)$. The second bubble is attached at the first one.

Example 3.5. In E as in the previous example consider the sections f_n given by

$$f_n(z) = \left(z, \frac{1}{n^2(z - \frac{1}{n})}, \frac{1}{n^2(z + \frac{1}{n})} \right).$$

Again they converge outside 0 in C^∞ to $f_\infty(z) = (z, 0, 0)$. As $n \to \infty$ the sequence (f_n) converges weakly to a cusp curve $\bar{f}$ and at 0 two bubbles $\{0\} \times S^2 \times \{0\}$ and $\{0\} \times \{0\} \times S^2$ arise and both are attached at $f_\infty(0)$. Note that the image of $\bar{f}$, that is

$$S^2 \times \{0\} \times \{0\} \;\cup\; \{0\} \times S^2 \times \{0\} \;\cup\; \{0\} \times \{0\} \times S^2 \subset E,$$

is not a singular surface as defined in Section 1. Indeed, there are *three* spheres glued together at *one* single point. However, it can be parametrized by a singular surface consisting of four spheres glued together as in Figure 23 such that $\bar{f}$ is constant on the middle sphere.

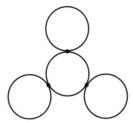

Figure 23.

The reader will obtain this singular surface by proceeding as suggested above.

Chapter VI

The squeezing theorem

This chapter describes an application of pseudo-holomorphic curves and its compactness properties in symplectic geometry, namely Gromov's famous squeezing theorem. Already proved in [Gr], it is among the first applications of pseudo-holomorphic curves at all. Gromov's proof of this result is based on an existence result for pseudo-holomorphic curves using methods from global analysis and Fredholm theory. It is far beyond the scope of this book to present these methods.

However, in Section 3 we give a survey on the relevant results. Using these results as 'black boxes', the reader who is not familiar with those methods should see how the existence proof works in principle and how crucial the compactness theorem is to that end. Then he or she might be encouraged to study the analytical setup in the literature and then more of the numerous applications of pseudo-holomorphic curves since [Gr]. To begin with, we recommend [ABK] and [MS2]. For further studies see also the references included in this book.

1. Discussion of the statement

The goal of this section is to state the squeezing theorem and discuss some of its consequences for symplectic geometry. The reader will find more on the squeezing theorem in textbooks on symplectic geometry (see for instance [HZ] or [MS2]).

Recall that a symplectic manifold (M, ω) is a manifold M of even dimension $2m$ together with a non-degenerate closed 2-form ω. A *volume form* on M is a differential form σ on M of degree $\dim M$ which is non-zero at each point $p \in M$.

Observe that ω being non-degenerate implies that

$$\omega^m := \underbrace{\omega \wedge \cdots \wedge \omega}_{m \text{ times}}$$

is a volume form on M and ω^m is called *the volume form of* (M, ω).

Consider now two symplectic manifolds (M, ω) and (M', ω'). A map $\phi: M \to M'$ is called *symplectic* if $\phi^* \omega' = \omega$. The symplectic manifolds (M, ω) and (M', ω') are called *symplectomorphic* if there exists a diffeomorphism $\psi: (M, \omega) \to (M', \omega')$ which is symplectic. Such a diffeomorphism is called a *symplectomorphism*.

In particular, each symplectomorphism preserves the corresponding volume forms, or in other words, is orientation and volume preserving. Consequently, the total volume is a symplectic invariant.

The total volume is the only invariant for volume preserving diffeomorphisms between compact oriented manifolds by the proposition below which is due to J. Moser [Ms] for closed manifolds and to D.V. Anosov & A.B. Katok [AK] and A. Banyaga [Ba] for compact manifolds with boundary.

Proposition 1.1. *Let N be a compact, connected and oriented manifold, possibly with boundary. Suppose σ and σ' are two volume forms on N with the same total volume, i.e.,*

$$\int_N \sigma = \int_N \sigma' .$$

Then there exists a diffeomorphism $\psi: (N, \sigma) \to (N, \sigma')$ with $\psi^* \sigma' = \sigma$, i.e., ψ is orientation and volume preserving.

In Riemannian geometry, for instance, curvature is a local invariant. In contrast, there are no local invariants in symplectic geometry which makes it more difficult to find invariants. The reason is

Proposition 1.2 (Darboux's theorem). *For each point p in a symplectic manifold (M, ω) there is a neighbourhood U of p in M such that $(U, \omega|_U)$ is symplectomorphic to an open subset in $(\mathbb{R}^{\dim M}, \omega_0)$.*

The proofs of Proposition 1.1 and Darboux's theorem are very similar, they are both based on Moser's method in [Ms], (see for instance, [HZ]).

Obvious symplectic invariants are the total volume, as mentioned before, and also the cohomology class of the symplectic structure. Based on his squeezing theorem, Gromov supplied in [Gr] the first non-obvious symplectic invariant, namely the symplectic radius.

Let $(x^1, y^1, \ldots, x^m, y^m)$ denote the standard coordinates in the Euclidean space $\mathbb{R}^{2m}$ together with its standard symplectic structure $\omega_0 = dx^1 \wedge dy^1 + \cdots + dx^m \wedge dy^m$ and its standard Euclidean scalar product. For $r > 0$ and each integer $m \geq 1$ we denote by $B^{2m}(r)$ and $\overline{B}^{2m}(r)$ the open and closed ball of radius r in $\mathbb{R}^{2m}$, respectively, centred at the origin. By $Z^{2m}(r)$ we denote the cylinder

$$Z^{2m}(r) := \left\{ (x^1, y^1, \ldots, x^m, y^m) \mid (x^1)^2 + (y^1)^2 < r^2 \right\} = B^2(r) \times \mathbb{R}^{2m-2} \subset \mathbb{R}^{2m} .$$

As submanifolds of $\mathbb{R}^{2m}$ these sets carry a standard symplectic structure.

Theorem 1.3 (Squeezing theorem for cylinders). *If there exists a symplectic embedding $\iota: \overline{B}^{2m}(r) \hookrightarrow Z^{2m}(R)$ then $r < R$.*

Remark. There is another proof of this result due to I. Ekeland and H. Hofer using Hamiltonian systems, (refer to [EH] and [HZ]).

Observe that for $r < R$ the inclusion $\overline{B}^{2m}(r) \subset \mathbb{R}^{2m}$ provides such an embedding.

The *symplectic radius* of a $2m$-dimensional symplectic manifold (M, ω) is defined as

$$\mathrm{symprad}(M, \omega) := \sup\left\{ r > 0 \mid \text{there is a symplectic embedding } \overline{B}_r^{2m} \hookrightarrow (M, \omega) \right\} .$$

Example 1.4 (see [Gr]). For $0 < r \leq R$ the symplectic radius of the symplectic product manifold $B^2(r) \times B^2(R)$ is clearly greater or equal to r and, since $B^2(r) \times B^2(R) \hookrightarrow Z^4(r)$, it is equal to r by the squeezing theorem. The volume of $B^2(r) \times B^2(R)$ is cr^2R^2, where $c > 0$ is a constant, and thus $B^2(r) \times B^2(R)$ is symplectomorphic to $B^2(r') \times B^2(R')$ if and only if $\{r, R\} = \{r', R'\}$.

Under the assumptions of the squeezing theorem, we can pick a basis $(v_1, \ldots, v_{2m-2})$ of $\mathbb{R}^{2m-2}$ such that $\iota(\overline{B}^{2m}(r))$ is contained in a fundamental domain of the canonical action of the group $\Gamma := \{n_1 v_1 + \cdots + n_{2m-2} v_{2m-2} \mid n_i \in \mathbb{Z} \text{ for } i = 1, \ldots, 2m - 2\}$ on the manifold $B^2(R) \times \mathbb{R}^{2m-2}$. Therefore we obtain a symplectic embedding

$$\text{pr} \circ \iota \colon \overline{B}^{2m}(r) \hookrightarrow B^2(R) \times T^{2m-2}.$$

Here T^{2m-2} denotes the torus $\Gamma \backslash \mathbb{R}^{2m-2}$ which inherits a standard symplectic structure from $\mathbb{R}^{2m-2}$ and pr: $B^2(R) \times \mathbb{R}^{2m-2} \rightarrow B^2(R) \times T^{2m-2}$ is the canonical projection. The claim of the squeezing theorem is now that in this situation necessarily $r < R$.

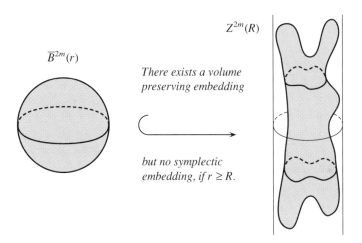

$Z^{2m}(R)$

$\overline{B}^{2m}(r)$

There exists a volume preserving embedding

but no symplectic embedding, if $r \geq R$.

Figure 24.

Given a symplectic manifold M of dimension $2m - 2$, it can be asked more generally under which conditions a symplectic embedding $\overline{B}^{2m}(r) \hookrightarrow B^2(R) \times M$ exists. Using pseudo-holomorphic curves, the following result can be proved.

Theorem 1.5 (Squeezing theorem). *Let M be a closed symplectic manifold of dimension $2m - 2$ and assume that the second fundamental group $\pi_2(M)$ is trivial. If there exists a symplectic embedding*

$$\iota \colon \overline{B}^{2m}(r) \hookrightarrow B^2(R) \times M$$

into the symplectic product $B^2(R) \times M$, then $r < R$.

Remark. The assumption $\pi_2(M) = \{0\}$ means that each continuous map $S^2 \to M$ is homotopic to a constant map and it guarantees that the theory of pseudo-holomorphic curves is applicable. Observe also that $\pi_2(T^{2m-2}) = \{0\}$.

In fact, using pseudo-holomorphic curves, the same result can be proved under a weaker assumption on the topology of (M, ω) (see [LM]). And, using other techniques, F. Lalonde and D. McDuff proved in [LM] the squeezing theorem without any additional assumption on the symplectic manifold (M, ω).

2. Proof modulo existence result for pseudo-holomorphic curves

In the present section we first state an existence result for J-holomorphic curves due to Gromov. Using this, we describe his proof of the squeezing theorem and postpone the discussion of the existence result to the next sections.

Let (M, ω) be a closed symplectic manifold with $\pi_2(M) = \{0\}$ and consider the symplectic product

$$(V, \Omega) := (S^2 \times M, \sigma \oplus \omega)$$

where σ is some symplectic structure on S^2. We denote by A the homology class in $H_2(V, \mathbb{Z})$ represented by a standard inclusion $S^2 \ni z \mapsto (z, q_0) \in S^2 \times M = V$.

Given a smooth map $\varphi \colon S^2 \to V$ we denote by $[\varphi]$ its homology class in $H_2(V, \mathbb{Z})$. Note that

$$\int_{S^2} \varphi^* \Omega =: \langle \Omega, [\varphi] \rangle$$

depends only on $[\varphi]$ by Stokes' theorem.

Proposition 2.1. *For each Ω-compatible almost complex structure J_V on V and any given point $p_0 \in V$ there exists a rational J_V-holomorphic curve $f \colon S^2 \to V$ in the homology class A containing p_0 in its image.*

Remarks. 1. The statement is trivial in case that J splits as $J = j \oplus J_M$. But in general, it is based on deep results from analysis and on the compactness theorem (see the next sections and the references there).

2. The choice of the homology class A will be crucial in the proof of the proposition, see the proof of Corollary 4.2 below. The same result is also true if J_V is only tamed by Ω. Existence results of this type are stated in [Gr].

In order to prove the squeezing theorem 1.5 assume now that we have a symplectic embedding

$$\iota \colon \overline{B}^{2m}(r) \hookrightarrow B^2(R) \times M \,.$$

We choose $\varepsilon \in (0, r)$ and a symplectic structure σ on S^2 such that the area $\mathscr{A}(S^2, \sigma)$ is equal to $\pi R^2 + \varepsilon$. Since $\mathscr{A}(B^2(R)) = \pi R^2 < \mathscr{A}(S^2, \sigma)$, we can view $B^2(R)$ as a subset of (S^2, σ) via a symplectic (i.e., volume preserving) embedding $B^2(R) \hookrightarrow S^2$ and hence ι as a symplectic embedding

$$\overline{B}^{2m}(r) \hookrightarrow (V, \Omega) = (S^2 \times M, \sigma \oplus \omega) \,.$$

Claim 2.2. There exists an Ω-compatible almost complex structure J on (V, Ω) such that

$$J|_{\iota(\overline{B}^{2m}(r-\varepsilon))} = \iota_*(J_0|_{\overline{B}^{2m}(r-\varepsilon)})$$

where J_0 is the standard complex structure on $\mathbb{R}^{2m}$.

Proof. Choose a Riemannian metric g on V such that

$$g|_{\iota(\overline{B}^{2m}(r-\varepsilon))} = \iota_*(g_0|_{\overline{B}^{2m}(r-\varepsilon)}) \tag{2.1}$$

and see the remark below how to obtain it. Now use the construction from Section I.2 (see the proofs of the Propositions I.2.2 and I.2.4) in order to get an almost complex structure J on V from the data Ω and g. Observe that this is a pointwise, unique construction and that, starting with the standard structures ω_0 and g_0 on $\mathbb{R}^{2m}$, the result is the standard complex structure J_0 on $\mathbb{R}^{2m}$. Consequently, J has the desired properties. $\qquad\square$

Remark. To construct a metric g with the property (2.1) a partition of unity argument can be used, since a positive linear combination of scalar products is again a scalar product.

Now apply Proposition 2.1 in order to choose a J-holomorphic curve $f: S^2 \to V$ representing A and containing $\iota(0)$ in its image. We claim that

$$\pi(r-\varepsilon)^2 \le \mathscr{A}\left(f(S^2) \cap \iota(\overline{B}^{2m}(r-\varepsilon))\right). \tag{2.2}$$

To that end note that $\iota(\overline{B}^{2m}(r-\varepsilon)) = \overline{B}_{r-\varepsilon}(\iota(0)) \subset (V, \Omega, J, g)$ is holomorphically isometric to the Euclidean ball $\overline{B}^{2m}(r-\varepsilon) \subset (\mathbb{R}^{2m}, J_0, g_0)$ by the choice of J and with $g := \Omega \circ (\mathbf{1} \times J)$. In particular, for each compact surface $S \subset S^2$ with boundary and $f(S) \subset \overline{B}^{2m}(r-\varepsilon)$, the holomorphic map $f|_S$ is absolutely area minimizing among all smooth maps $\varphi: S \to \overline{B}^{2m}(r-\varepsilon)$ with $\varphi|_{\partial S} = f|_{\partial S}$ (see Section II.2). Observe also that the image of f is not contained in $\iota(\overline{B}^{2m}(r))$ for otherwise f would be constant.

The inequality (2.2) is the classical monotonicity lemma for minimal surfaces in Euclidean space and its proof can be easily recovered from the arguments in Section II.1. Indeed, in order to show that the statement of the monotonicity lemma II.1.3 holds for area minimizing maps from compact surfaces with boundary to some Euclidean space with $C_{\mathrm{ML}} = \pi$ and $\varepsilon_0 = \infty$ just copy the proof of the monotonicity lemma II.1.3 but use Proposition A.1 in the appendix instead of Lemma II.3.1.

In order to prove now that $r < R$ continue inequality (2.2) as follows:

$$\pi(r-\varepsilon)^2 \le \mathscr{A}\left(f(S^2) \cap \iota(\overline{B}^{2m}(r-\varepsilon))\right) < \mathscr{A}(f) = \int_{S^2} f^*\Omega = \int_{S^2} \mathrm{incl}^*\Omega = \pi R^2 + \varepsilon$$

where incl: $S^2 \hookrightarrow S^2 \times M$ denotes a standard inclusion. The first and second equality follow from the fact that f is J-holomorphic with Ω-compatible J and that f represents $A \in H_2(V, \mathbb{R})$, respectively. Thus it follows that $r \le R$ and then clearly $r < R$ since $\iota(\overline{B}^{2m}(r)) \subset B^2(R) \times M$ is compact.

This finishes the proof of the squeezing theorem up to the existence result stated in Proposition 2.1 which the remainder of this chapter is devoted to.

3. The analytical setup: A rough outline

This section contains a survey on the analytic methods used to show that for any almost complex structure J on V there are sufficiently many elements in the set

$$M(A,J) := \Big\{ f\colon S^2 \to V \mid f \text{ is } J\text{-holomorphic}, [f] = A \Big\}$$

in order to prove Proposition 2.1. To that end, one writes $M(A,J)$ as the zero set of the *Cauchy-Riemann operator* $\bar{\partial}_J$ in the following way: Put

$$\mathcal{B} := \Big\{ f \in C^\infty(S^2, V) \mid [f] = A \Big\}$$

and define for $f \in \mathcal{B}$ the vector space

$$\mathcal{E}_{(f,J)} := \Big\{ \alpha \mid \alpha \text{ is an anti-}J\text{-linear smooth section of } \mathrm{Hom}_{\mathbb{R}}(TS^2, f^*TV) \to S^2 \Big\}.$$

Here $f^*TV \to S^2$ is the pull back bundle of $TV \to V$ via f. The fibre of the bundle $\mathrm{Hom}_{\mathbb{R}}(TS^2, f^*TV) \to S^2$ over $s \in S^2$ is $\mathrm{Hom}_{\mathbb{R}}(T_sS^2, T_{f(s)}V)$. Set-theoretically, we obtain a bundle $\mathcal{E}_J \to \mathcal{B}$ with fibre $\mathcal{E}_{(f,J)}$ over $f \in \mathcal{B}$. For $f \in \mathcal{B}$ there is a unique decomposition of the differential Tf into a J-linear and an anti-J-linear part, namely, $Tf = (1/2)(Tf - J \circ Tf \circ j) + (1/2)(Tf + J \circ Tf \circ j)$. The Cauchy-Riemann operator picks the anti-J-linear part of it, that is

$$\bar{\partial}_J(f) := \frac{1}{2}\left(Tf + J \circ Tf \circ j\right).$$

Observe now that $\bar{\partial}_J\colon f \mapsto \bar{\partial}_J(f)$ is a section of $\mathcal{E}_J \to \mathcal{B}$ and that

$$M(A,J) = \bar{\partial}_J^{-1}(0)$$

is the intersection of $\bar{\partial}_J$ with the zero section.

Notice that the spaces $\mathcal{E}_J$ and $\mathcal{B}$ are not finite-dimensional. However, let us first consider the analogous situation of a finite-dimensional vector bundle $E \to B$. The zero set $s^{-1}(0)$ of a smooth section s of $E \to B$ is a manifold if s is transverse to the zero section. We describe this now in more detail.

The finite-dimensional analog. Suppose X, Y and Z are manifolds. Smooth maps $\varphi\colon X \to Z$ and $\psi\colon Y \to Z$ are called *transverse* if for any $(x, y) \in X \times Y$ with $\varphi(x) = \psi(y) =: z$ the subspaces $T_x\varphi(T_xX)$ and $T_y\psi(T_yY)$ span T_zZ. If, moreover, ψ is an embedding it is also said that the submanifold $V := \psi(Y)$ is *transverse* to φ; this condition is clearly independent of the choice of the embedding ψ.

If the submanifold $V \subset Z$ is transverse to φ, then $W := \varphi^{-1}(V)$ is a submanifold of X with $\mathrm{codim}_X W = \mathrm{codim}_Z V$ or $W = \varnothing$ (see [Hi], Chapter 1, Theorem 3.3). If X is a manifold with boundary, $V \subset Z$ is a submanifold and both, φ and $\varphi|_{\partial X}$ are transverse to V, then $W = \varphi^{-1}(V)$ is a submanifold of X with boundary. Furthermore, $\partial W = W \cap \partial X$ (see [Hi], Chapter 1, Theorem 4.2).

Recall also that the previous results follow immediately form the inverse function theorem.

In this context we want to describe another classical result, namely a theorem of A. Sard. Let $\varphi \colon X \to Z$ be a smooth map between finite-dimensional manifolds. A point $x \in X$ is called a *regular point* of φ if $T_x\varphi \colon T_xX \to T_{\varphi(x)}Z$ is surjective and $z \in Z$ is called a *regular value* of φ if each $x \in \varphi^{-1}(z)$ is a regular point of φ. Elements of X (respectively Z) which are not regular points (respectively regular values) are called *singular points* (respectively *singular values*) of φ. Observe that each $z \in Z \setminus \varphi(X)$ is by definition a regular value of φ. A subset of a topological space is called *residual* if it contains a countable intersection of open and dense subsets. By Baire's theorem, residual subsets of complete metric spaces are dense.

Proposition 3.1 (Sard's theorem). *Let $\varphi \colon X \to Z$ be a smooth map between finite-dimensional manifolds. Then the set of regular values of φ is residual in Z.*

Remark. Actually Sard's theorem states more. It is only required that φ is of class C^r with $r > \min\{0, \dim X - \dim Z\}$ and it states that the set of singular values has measure zero. The differentiability assumption is optimal. See [Hi] or [Mi] for details.

In the sequel, we say *generic* instead of "for a residual set" and from the context it will be clear which subset of which topological space we mean.

From Sard's theorem it follows that generically, $\varphi^{-1}(z)$ is either a smooth submanifold of X with $\operatorname{codim}_X \varphi^{-1}(z) = \dim Z$ or it is empty.

Using Sard's theorem it can be proved that, given $\varphi \colon X \to Z$, the set of smooth embeddings $\psi \colon Y \to Z$ which are transverse to φ is residual in the set of smooth embeddings $Y \hookrightarrow Z$ with respect to the C^∞-topology. For details, we refer to Chapter 3 of [Hi].

Let us now come back to the original problem, namely investigating the zero set of the section $\bar{\partial}_J$ of $\mathcal{E}_J \to \mathcal{B}$. To that end one uses infinite-dimensional versions of the results described before. Such versions exist for Banach manifolds. However, the spaces $\mathcal{E}_J$ and $\mathcal{B}$ are not Banach manifolds, but after enlarging these spaces by weakening the differentiability conditions on their elements, they are. It will turn out that for a generic choice of J, the section $\bar{\partial}_J$ is transverse to the zero section and that the manifold $\bar{\partial}_J^{-1}(0)$ is finite-dimensional or empty.

Banach manifolds and Fredholm maps. The notion of differentiability for maps between Banach spaces is essentially the same as in finite dimensions. Let $\mathcal{U}$ and $\mathcal{V}$ be open subsets of Banach spaces $(\mathcal{F}, \| \cdot \|)$ and $(\mathcal{G}, \| \cdot \|)$, respectively. Then a map $\varphi \colon \mathcal{U} \to \mathcal{V}$ is called *differentiable* at $u \in \mathcal{U}$ if there exists a *continuous* linear map $Q \colon \mathcal{F} \to \mathcal{G}$ such that the map ε, defined on a neighbourhood of zero in $\mathcal{F}$ and given by $\varphi(u + h) = \varphi(u) + Q(h) + \|h\|\varepsilon(h)$ and $\varepsilon(0) = 0$, is continuous. Then Q is called the *differential* of φ at u.

The definition of a Banach manifold is analogous to the definition of a finite-dimensional manifold. A *Banach manifold* $\mathcal{M}$ is a separable Hausdorff space modeled on a separable Banach space instead of some $\mathbb{R}^n$ in the finite-dimensional case. By the above, the notions of *tangent space, tangent map, C^k-map, diffeomorphism, regular value and regular points,...* are clearly defined in this setting.

Important is that the *inverse function theorem* extends to differentiable maps between Banach manifolds. For instance, the corollary below can be proved in the same way as in the finite-dimensional case. However, with inverse function theorem arguments one has to be aware that a closed linear subspaces $\mathcal{F}_1$ of a Banach space $\mathcal{F}$ in general does not have a topological complement. We recall that a linear subspace $\mathcal{F}_1$ of a Banach space $\mathcal{F}$ is said to have a *topological complement* if there exists a Banach space $\mathcal{F}_2$ together with a continuous linear isomorphism $\mathcal{F}_1 \times \mathcal{F}_2 \to \mathcal{F}$. This is equivalent to saying that there exists a closed linear subspace $\mathcal{F}_2$ of $\mathcal{F}$, complementary to $\mathcal{F}_1$. Then it is also said that $\mathcal{F}_1 \subset \mathcal{F}$ *splits*.

Corollary 3.2 (of the inverse function theorem). *Suppose $\varphi\colon \mathcal{X} \to \mathcal{Z}$ is a smooth map between Banach manifolds and that $T_x\varphi$ is surjective. Assume furthermore that $\ker T_x\varphi$ splits in $T_x\mathcal{X}$. Then there exists an open neighbourhood $\mathcal{W}$ of $\varphi(z)$ in $\mathcal{Z}$, an open subset $\mathcal{V}$ in $\ker T_x\mathcal{X}$ and a diffeomorphism $\xi\colon \mathcal{U} \to \mathcal{V} \times \mathcal{W}$ from an open neighbourhood $\mathcal{U}$ of x in $\mathcal{X}$ such that*

$$\varphi \circ \xi^{-1} \colon \mathcal{V} \times \mathcal{W} \to \mathcal{W}$$

is just the projection onto the second factor.

For more details on the objects and results described so far in this paragraph we refer the reader to [Lg1] and [Lg2].

A *Fredholm operator* is a continuous linear map $L\colon \mathcal{F} \to \mathcal{G}$ between Banach spaces such that the kernel of L is finite-dimensional, the image $L(\mathcal{F})$ is closed and has finite codimension in $\mathcal{G}$. The *index* of L is defined to be

$$\operatorname{index} L := \dim \ker L - \dim \operatorname{coker} L\,,$$

i.e., the difference of the dimension of the kernel and codimension of the image of L. An important fact is that the set $F(\mathcal{F}, \mathcal{G})$ of Fredholm operators from $\mathcal{F}$ to $\mathcal{G}$ is open in the Banach space $L(\mathcal{F}, \mathcal{G})$ of continuous linear maps from $\mathcal{F}$ to $\mathcal{G}$. Furthermore, the index is continuous on $F(\mathcal{F}, \mathcal{G})$ and, having a discrete image, it is constant on the connected components of $F(\mathcal{F}, \mathcal{G})$. For proofs the reader may consult [Lg1].

A C^1-map $\varphi\colon \mathcal{X} \to \mathcal{Z}$ from one Banach manifold to another is called a *Fredholm map* if at each point $x \in \mathcal{X}$ its tangent map $T_x\varphi\colon T_x\mathcal{X} \to T_{\varphi(x)}\mathcal{Z}$ is a Fredholm operator. By the above, if $\mathcal{X}$ is connected the index of $T_x\varphi$ does not depend on the choice of $x \in \mathcal{X}$ and it is then called the *index* of φ.

Observe that a finite-dimensional subspace of a Banach space has always a topological complement by the theorem of Hahn-Banach. Thus, if $z \in \mathcal{Z}$ is a regular value of a smooth Fredholm map $\varphi\colon \mathcal{X} \to \mathcal{Z}$ then $\varphi^{-1}(z)$ is a finite-dimensional submanifold of $\mathscr{X}$ with $\dim \varphi^{-1}(z) = \text{index } \varphi$. Furthermore, if $W \subset \mathscr{Z}$ is a finite-dimensional sub-manifold transverse to φ, then $\varphi^{-1}(W)$ is a smooth submanifold of $\mathscr{X}$ of dimension index $\varphi + \dim W$. The generalization of Sard's theorem and the transversality result are due to S. Smale [Sm].

Proposition 3.3 (Sard-Smale theorem). *The set of regular values of a smooth Fredholm map $\mathscr{X} \to \mathscr{Z}$ is residual in $\mathscr{Z}$.*

Proposition 3.4 (Transversality theorem). *Let $\varphi \colon \mathscr{X} \to \mathscr{Z}$ be a smooth Fredholm map and $\psi \colon Y \to \mathscr{Z}$ be an embedding of a finite-dimensional manifold Y into $\mathscr{Z}$. Suppose φ is transverse to ψ on a closed subset W of Y. Then there exists a map $\psi' \colon Y \to \mathscr{Z}$, arbitrary close to ψ in the C^1-topology such that ψ' is transverse to φ and $\psi'|_W = \psi$.*

For the proofs of these results we refer to the original paper [Sm] by Smale. The proof of the Sard-Smale theorem consists of a reduction to the finite-dimensional statement and it suffices to assume that the Fredholm map in question is of class C^r with r greater than its index and positive.

The moduli space of pseudo-holomorphic curves. Next we briefly describe the analytical setup in our situation. The reader will find details in a more general situation in [ABK] or [MS2].

We may assume that V is embedded into $\mathbb{R}^N$ for a sufficiently large integer $N > 0$. For $p > 2$ the *Sobolev space* $W^{1,p}(S^2, \mathbb{R}^N)$ is the completion of $C^\infty(S^2, \mathbb{R}^N)$ with respect to the norm

$$\|f\|_{1,p} := \left(\int_{S^2} \left(|f|^p + |Df|^p \right) \sigma \right)^{1/p}.$$

After fixing a Riemannian metric on S^2, one may define σ to be the induced volume form on S^2 and the norms inside the integral are the canonical norms on the corresponding finite-dimensional vector space.

The normed vector space $(W^{1,p}(S^2, \mathbb{R}^N), \| \cdot \|_{1,p})$ is a Banach space and it embeds naturally into $C^0(S^2, \mathbb{R}^N)$. We put

$$\mathscr{B}^{1,p} := \left\{ f \in W^{1,p}(S^2, \mathbb{R}^N) \mid f(S^2) \subset V, [f] = A \right\}$$

and it can be shown that $\mathscr{B}^{1,p}$ is a Banach manifold for $p > 2$.

The differential of $f \in \mathscr{B}^{1,p}$ is in L^p, that is to say, $\int_{S^2} |Df|^p \sigma < \infty$ and it defines an L^p-section of the vector bundle $\text{Hom}_{\mathbb{R}}(TS^2, f^*TV) \to S^2$. Define a Banach space

$$\mathscr{E}^p_{(f,J)} := \left\{ \alpha \mid \alpha \text{ is an anti-}J\text{-linear } L^p\text{-section of } \text{Hom}_{\mathbb{R}}(TS^2, f^*TV) \to S^2 \right\}$$

with $\|\alpha\|_p := \left(\int_{S^2} |\alpha|^p \sigma \right)^{1/p}$ for $\alpha \in \mathscr{E}^p_{(f,J)}$. Then $\mathscr{E}^p_J \to \mathscr{B}^{1,p}$ inherits a structure of a Banach bundle with fibre $\mathscr{E}^p_{(f,J)}$ over $f \in \mathscr{B}^{1,p}$. The Cauchy-Riemann operator $\bar{\partial}_J$ defines a smooth section of $\mathscr{E}^p_J \to \mathscr{B}^{1,p}$. The tangent space to $\mathscr{E}^p_J$ at a point 0_f in the zero section splits naturally as $T_{0_f} \mathcal{E}_J^p \simeq T_f \mathcal{B}^{1,p} \times \mathcal{E}_{(f,J)}^p$. Suppose now that $\bar{\partial}_J(f) = 0$ for some $f \in \mathcal{B}^{1,p}$. Then we denote by

$$D\bar{\partial}_J(f): T_f \mathcal{B}^{1,p} \to \mathcal{E}_{(f,J)}^p$$

the tangent map of $\bar{\partial}_J$ at f composed with the canonical projection $T_{0_f} \mathcal{E}_J^p \to \mathcal{E}_{(f,J)}^p$. It can be shown that $\bar{\partial}_J$ is an elliptic differential operator and using this one obtains

Proposition 3.5. *Suppose that $f \in \mathcal{B}^{1,p}$ and $\bar{\partial}_J(f) = 0$. Then $D\bar{\partial}_J(f)$ is a Fredholm operator and f is smooth.*

The Sard-Smale theorem gives hope that, given J there is an almost complex structure J' on V which is a small perturbation of J such that $D\bar{\partial}_{J'}(f)$ is surjective. In order to show this, one has to consider variations of J in a suitable space $\mathcal{J}$ of almost complex structures. Again, there is the problem that the space $\mathcal{J}_c$ of smooth almost complex structures on V compatible with Ω is not a Banach manifold. But there exists a subset $\mathcal{J} \subset \mathcal{J}_c$ which is dense in the C^∞-topology and which can be given the structure of a Banach manifold such that the inclusion $\mathcal{J} \hookrightarrow \mathcal{J}_c$ is continuous (see [ABK]). An almost complex structure $J \in \mathcal{J}$ is called *regular* if $D\bar{\partial}_J(f)$ is surjective for every $f \in \bar{\partial}_J^{-1}(0)$ and we denote by

$$\mathcal{J}_{\mathrm{reg}} := \left\{ J \in \mathcal{J} \mid J \text{ is regular} \right\}$$

the set of regular almost complex structures in $\mathcal{J}$.

We consider now the so called *universal moduli space* of pseudo-holomorphic curves in V representing A which is the topological subspace

$$\mathcal{M}(A, \mathcal{J}) := \left\{ (f, J) \in \mathcal{B}^{1,p} \times \mathcal{J} \mid \bar{\partial}_J(f) = 0 \right\}.$$

of $\mathcal{B}^{1,p} \times \mathcal{J}$. By Proposition 3.5, f is smooth for each $(f, J) \in \mathcal{M}(A, \mathcal{J})$.

Remark 3.6. We mentioned before that $\mathcal{B}^{1,p}$ embeds into $C^0(S^2, \mathbb{R}^N)$. Thus, a convergent sequence $(f_n, J_n)_{n \geq 1}$ in $\mathcal{M}(A, \mathcal{J})$ is also convergent in C^∞ by the generalized Weierstraß theorem III.4.1. Conversely, a sequence $(f_n, J_n)_{n \geq 1}$ in $\mathcal{M}(A, \mathcal{J})$ which converges in $\mathcal{B} \times \mathcal{J}$ converges clearly in $\mathcal{B}^{1,p} \times \mathcal{J}$.

Consequently, $\mathcal{M}(A, \mathcal{J})$ is also a topological subspace of $\mathcal{B} \times \mathcal{J}$ where $\mathcal{B}$ carries the C^∞-topology. Summing up, by enlarging the space $\mathcal{B}$ to $\mathcal{B}^{1,p}$ in order that the analysis works, the solution space of the equation $\bar{\partial}_J(\cdot) = 0, J \in \mathcal{J}$, does not change.

The projection $\mathcal{M}(A, \mathcal{J}) \ni (f, J) \mapsto J$ is in the sequel denoted by

$$P: \mathcal{M}(A, \mathcal{J}) \to \mathcal{J}$$

and we observe that $P^{-1}(J) = M(A, J) \times \{J\}$. The results based on methods from global analysis and the theory of the Cauchy-Riemann operator which go into the proof of Proposition 2.1 are summarized in the following

Proposition 3.7. *The universal moduli space $\mathcal{M}(A, \mathcal{J})$ is a Banach manifold and the projection $P\colon \mathcal{M}(A, \mathcal{J}) \to \mathcal{J}$ is a Fredholm map. The set of regular values of P is $\mathcal{J}_{\mathrm{reg}}$. In particular, $\mathcal{J}_{\mathrm{reg}} \subset \mathcal{J}$ is dense.*

Remarks. Notice that this proposition shows that for a generic J the moduli space $M(A, J)$ is a manifold of dimension index $P = $ index $D\bar{\partial}_J(f)$ for $f \in M(A, J)$ or is empty. The index of $D\bar{\partial}_J(f)$ can be computed from topological data (see [ABK] or [MS2]) but we will not need this.

The major step for proving the proposition uses the theory of the Cauchy-Riemann operator (see [ABK] and [MS2]) which we do not describe here. Apart from that, we give an outline of the arguments:

One can construct a Banach bundle $\mathcal{E}^p \to \mathcal{B}^{1,p} \times \mathcal{J}$ with fibre $\mathcal{E}^p_{(f,J)}$ over $(f, J) \in \mathcal{B}^{1,p} \times \mathcal{J}$. Then

$$\bar{\partial}\colon \mathcal{B}^{1,p} \times \mathcal{J} \to \mathcal{E}^p, \quad (f, J) \mapsto \bar{\partial}_J(f)$$

is a smooth section of that bundle and $\mathcal{M}(A, \mathcal{J}) = \bar{\partial}^{-1}(0)$. For $(f, J) \in \bar{\partial}^{-1}(0)$ consider the map

$$D\bar{\partial}(f, J)\colon T_f\mathcal{B}^{1,p} \times T_J\mathcal{J} \to \mathcal{E}^p_{(f,J)}$$

which is defined to be the tangent map of $\bar{\partial}$ at (f, J) composed with the natural projection onto the fibre $\mathcal{E}^p_{(f,J)}$. We write $D\bar{\partial}(f, J)(\xi, \eta) = D_1\bar{\partial}(f, J)(\xi) + D_2\bar{\partial}(f, J)(\eta)$ for $(\xi, \eta) \in T_f\mathcal{B}^{1,p} \times T_J\mathcal{J}$. Observe that

$$D_1\bar{\partial}(f, J) = D\bar{\partial}_J(f).$$

In order to make the following considerations clearer, we abbreviate $D := D\bar{\partial}(f, J)$, $D_1 := D_1\bar{\partial}(f, J), D_2 := D_2\bar{\partial}(f, J), \mathcal{F} := T_f\mathcal{B}^{1,p}, \mathcal{G} := T_J\mathcal{J}, \mathcal{H} := \mathcal{E}_{(f,J)}$ and thus

$$D = (D_1, D_2)\colon \mathcal{F} \times \mathcal{G} \to \mathcal{H}.$$

The key argument for proving the proposition is that, using the theory of the Cauchy-Riemann operator, it can be shown that D is surjective (see [ABK] and [MS2]).

Since D_1 is Fredholm and D is surjective, a finite-dimensional subspace $G \subset \mathcal{G}$ can be chosen such that $D_2|_G$ is injective and its image $D_2(G)$ is a topological complement of $D_1(\mathcal{F})$. If $\mathcal{F}_1$ is a topological complement of $\ker D_1$ then clearly $\mathcal{F}_1 \times G$ is a topological complement of $\ker D$. Corollary 3.2 yields that $\mathcal{M}(A, \mathcal{J}) = \bar{\partial}^{-1}(0)$ is a submanifold of $\mathcal{B}^{1,p} \times \mathcal{J}$ with $T_{(f,J)}\mathcal{M}(A, \mathcal{J}) = \ker D$.

With the notation from above, the tangent map of P at (f, J) is

$$\Pi := T_{(f,J)}P\colon \ker D \to \mathcal{G}, \quad (\xi, \eta) \mapsto \eta$$

and it follows that $\ker \Pi = \ker D_1 \times \{0\}$ and is in particular finite-dimensional. Observe now that

$$\Pi(\ker D) = \left\{ \eta \in \mathcal{G} \mid D_2\eta \in D_1(\mathcal{F}) \right\}$$

and this is a closed subspace of $\mathcal{G}$. Furthermore,

$$\mathcal{G}/_{\Pi(\ker D)} \to \mathcal{H}/_{D_1(\mathcal{F})}, \quad \eta + \Pi(\ker D) \mapsto D_2\eta + D_1(\mathcal{F})$$

is well defined and injective. It is surjective since D is surjective. In particular, coker Π is finite-dimensional and thus P is a Fredholm map. Moreover, since

$$\operatorname{coker} T_{(f,J)}P = \mathcal{G}/_{\Pi(\ker D)} \simeq \mathcal{H}/_{D_1(\mathcal{F})} = \operatorname{coker} D\bar{\partial}_J(f)$$

and $\ker T_{(f,J)}P \simeq \ker D\bar{\partial}_J(f)$, we see that the regular values of P are precisely the elements in $\mathcal{J}_{\mathrm{reg}}$. The previous argument also shows that index $P = $ index $D\bar{\partial}_J(f)$.

4. The required existence result

This section is devoted to the proof of Proposition 2.1. Our presentation follows closely [LM] and other references are [ABK] and [MS2].

Let (M, ω) be a closed symplectic manifold with $\pi_2(M) = \{0\}$ and set $(V, \Omega) = (S^2 \times M, \sigma \oplus \omega)$ as before. Pick an arbitrary Ω-compatible almost complex structure J_V on V and some point

$$p_0 = (z_0, q_0) \in S^2 \times M.$$

The goal is to show that there exists a rational J_V-holomorphic curve $f: S^2 \to V$ in the homology class $A \in H_2(V, \mathbb{Z})$ represented by a standard embedding $S^2 \hookrightarrow S^2 \times M$.

Recall that $P: \mathcal{M}(A, \mathcal{J}) \to \mathcal{J}$ denotes the canonical projection which is Fredholm. We identify now $M(A, J)$ with $M(A, J) \times \{J\}$. Viewing $M(A, J)$ as a subset of $\mathcal{M}(A, \mathcal{J})$ we make a slight abuse of notation and write also $f \in \mathcal{M}(A, \mathcal{J})$ instead of $(f, J) \in \mathcal{M}(A, \mathcal{J})$ for $f \in M(A, J)$. This should not give rise to any confusions.

From now on we write $G := \operatorname{Conf}(S^2)$ for the Lie group of conformal transformations of the Riemann sphere (see Section I.4). If J_M is an almost complex structure on M then we have that

$$P^{-1}(j \oplus J_M) = \left\{ \sigma_q: z \mapsto (\sigma(z), q) \mid \sigma \in G, q \in M \right\}. \tag{4.1}$$

Indeed, composing $f \in P^{-1}(j \oplus J_M)$ with the projection $S^2 \times M \to M$, a J_M-holomorphic curve in M is obtained since $j \oplus J_M$ is a product structure. Since $\pi_2(M) = \{0\}$ this J_M-holomorphic curve is homotopic to a constant map and hence it is constant itself.

In the present situation, it turns out that regular points of P arise for a split almost complex structure which is somewhere integrable, (see Section 3.5 in [MS2]).

Fact 4.1. *Suppose that J_M is integrable on an open set $U_0 \subset M$. Then, for each $\sigma \in G$ and each $q \in U_0$ the $j \oplus J_M$-holomorphic curve σ_q is a regular point of P.*

Therefore, choose J_M such that it is integrable on an open neighbourhood U_0 of q_0 in M and we also require that J_M is compatible with ω. This is clearly possible by Darboux's theorem and the arguments in the proof of Claim 2.2.

There is a natural map $\mathcal{M}(A, \mathcal{J}) \times S^2 \to V$ given by the evaluation $(f, z) \mapsto f(z)$. This map factors through the action of G on $\mathcal{M}(A, \mathcal{J}) \times S^2$ given by

$$\sigma \cdot (f, z) = (f \circ \sigma^{-1}, \sigma(z)) \tag{4.2}$$

for $\sigma \in G$ and $(f, z) \in \mathcal{M}(A, \mathcal{J}) \times S^2$. We write $\mathcal{M}(A, \mathcal{J}) \times_G S^2$ for corresponding quotient space and $[f, z]$ for the equivalence class of (f, z). For each $J \in \mathcal{J}$ the set $M(A, J) \times S^2 \subset \mathcal{M}(A, \mathcal{J}) \times S^2$ is invariant under the G-action and we denote by

$$\mathrm{ev}_J \colon M(A, J) \times_G S^2 \to V$$

the resulting *evaluation map* on the corresponding quotient space. Assume now that $J \in \mathcal{J}_{\mathrm{reg}}$. It is readily verified that G acts properly and freely on the finite-dimensional *manifold* $M(A, J) \times S^2$. We recall that G *acting properly* means that for each compact subset K of $M(A, J) \times S^2$, the set $\{ \sigma \in G \mid \sigma(K) \cap K \neq \varnothing \}$ is a compact subset of G. It is standard and easy to see that G acting properly and freely implies that the Hausdorff space $M(A, J) \times_G S^2$ inherits a differentiable structure from $M(A, J) \times S^2$ such that the canonical projection is a submersion with the fibres being the G-orbits.

One of the main ingredients of the proof of Proposition 2.1 is the following

Corollary 4.2 (of the compactness theorem). *Suppose that $([f_n, z_n])_{n \geq 1}$ is a sequence in $\mathcal{M}(A, \mathcal{J}) \times_G S^2$ and that some subsequence of $(P(f_n))_{n \geq 1}$ converges in $\mathcal{J}$. Then $([f_n, z_n])$ has a convergent subsequence.*

Before proving the corollary, we make some comment on the choice of the homology class A. Since $\pi_2(M) = \{0\}$ by assumption, $\langle \Omega, A \rangle$ is the smallest positive value of $\langle \Omega, \cdot \rangle$ on spherical homology classes $\sum_{i=1}^k n_i [\varphi_i]$, $n_i \in \mathbb{Z}$ and $\varphi_i \colon S^2 \to V$. This is important in the sequel.

Proof of the Corollary. Let $([f_n, z_n])_{n \geq 1}$ be as in the assumption of the corollary. In other words, (f_n) is a sequence of J_n-holomorphic curves in $\mathcal{M}(A, \mathcal{J})$ where (J_n) has a convergent subsequence and $z_n \in S^2$. After passing to a subsequence, we may assume that (J_n) converges to some $\bar{J} \in \mathcal{J}$ and (z_n) to some $z \in S^2$.

The claim is that there exists a sequence (σ_n) in G such that $(f_n \circ \sigma_n)$ has a uniformly convergent subsequence.

Since each f_n represents A and J_n is Ω-compatible, the area of f_n is given by $\langle \Omega, A \rangle$. Thus the sequence (f_n) has bounded area. The compactness theorem implies that, after passing to a subsequence, (f_n) converges weakly to a $\bar{J}$-holomorphic cusp curve $\bar{f} \colon (\bar{S}, \bar{j}) \to V$ (see also the Remark at the end of Section V.2). Let $S_1, \ldots, S_k$ be the components of the singular surface $\bar{S}$ on which $\bar{f}$ is not constant. Clearly each S_i, $i = 1, \ldots, k$, is a sphere. If $k \geq 2$ then

$$0 < \langle \Omega, [\bar{f}|_{S_i}] \rangle = \mathcal{A}(\bar{f}|_{S_i}) < \mathcal{A}(\bar{f}) = \langle \Omega, A \rangle$$

in contradiction to the fact that $\langle \Omega, A \rangle$ is the smallest positive value of $\langle \Omega, \cdot \rangle$ on spherical homology classes. Hence $k = 1$ and we may take $(\bar{S}, \bar{j}) = (S^2, j)$.

By the definition of weak convergence, there are diffeomorphisms $\varphi_n \colon S^2 \to \bar{S}$ such that $(\varphi_n^{-1})^* j =: j_n$ converges in C^∞ to $\bar{j}$ and $f_n \circ \varphi_n^{-1}$ converges uniformly to $\bar{f}$. Choose now three points $s_1, s_2, s_3 \in \bar{S} = S^2$ and let $\psi_n \colon (S^2, j) \to (\bar{S}, j_n)$ be the unique biholomorphic transformation with $\psi_n(s_i) = s_i$ for $i = 1, 2, 3$. We put $\sigma_n := \varphi_n^{-1} \circ \psi_n$ and observe that $\sigma_n \in G$. Applying the compactness theorem it is readily seen that (ψ_n) converges uniformly to the identity and thus $(f_n \circ \sigma_n)$ converges uniformly to $\bar{f}$.   □

Lemma 4.3. *Given neighbourhoods $\mathcal{U}$ of $j \oplus J_M$ in $\mathcal{J}$ and U of p_0 in V there exist an almost complex structure $J_a \in \mathcal{U} \cap \mathcal{J}_{\mathrm{reg}}$ and some $p_a \in U$ such that p_a is a regular value of ev_{J_a} and $\mathrm{ev}_{J_a}^{-1}(p_a)$ consists of exactly one point.*

Proof. From the projection P and the evaluation we obtain a Fredholm map

$$\Phi \colon \mathcal{M}(A, \mathcal{J}) \times S^2 \to \mathcal{J} \times V, \quad (f, z) \mapsto (P(f), f(z)).$$

From (4.1) and Fact 4.1 we see that $(j \oplus J_M, p_0)$ is a regular value of Φ. Thus there is an open connected neighbourhood $\mathcal{V}$ of $(\mathbf{1}_{q_0}, z_0)$ in $\mathcal{M}(A, \mathcal{J}) \times S^2$ and an open neighbourhood $\mathcal{W}$ of $\Phi(\mathbf{1}_{q_0}, z_0) = (j \oplus J_M, p_0)$ in $\mathcal{J} \times V$ such that $\Phi|_{\mathcal{V}} \colon \mathcal{V} \to \mathcal{W}$ is a submersion. By Corollary 3.2, we may assume that there is a diffeomorphism $\xi \colon \mathcal{W} \times B \to \mathcal{V}$ with $B = \Phi|_{\mathcal{V}}^{-1}(j \oplus J_M, p_0)$ being a connected and open ball in some $\mathbb{R}^n$ such that

$$\Phi \circ \xi \colon \mathcal{W} \times B \to \mathcal{W}$$

is the projection onto $\mathcal{W}$.

Since modulo reparametrization there is exactly one $j \oplus J_M$-holomorphic curve through p_0 we see that $\dim B = \dim G$. Therefore, choosing $(w, b) \in \mathcal{W} \times B$, the set of points in $\{w\} \times B$ which are equivalent to (w, b) modulo G is open in $\{w\} \times B$. Since it is evidently a closed subset of $\{w\} \times B$ it is equal to $\{w\} \times B$. Thus, given $(J, p) \in \mathcal{W}$ there is modulo G a unique J-curve in $\mathcal{V}$ passing through p. It follows from the Sard-Smale theorem that there exists a sequence (J_n) in $\mathcal{J}_{\mathrm{reg}}$ converging to $j \oplus J_M$ and elements $(f_n, z_n) \in M(A, J_n) \times S^2$ converging to $(\mathbf{1}_{q_0}, z_0)$ such that $[f_n, z_n]$ is a regular point of ev_{J_n}.

If the lemma were not true, we may assume there exist elements $(f_n', z_n') \in M(A, J_n) \times S^2$ with $[f_n', z_n']$ being a singular point of ev_{J_n} and $\mathrm{ev}_{J_n}[f_n', z_n'] = \mathrm{ev}_{J_n}[f_n, z_n]$. By Corollary 4.2 we may further assume that (f_n', z_n') converges to some $(f', z') \in M(A, j \oplus J_M) \times S^2$. Since $f'(z') = \lim f_n'(z_n') = \lim f_n(z_n) = p_0$ and since there is modulo G exactly one $j \oplus J_M$-holomorphic curve in $M(A, j \oplus J_M)$ passing through p_0 we can suppose that $(f', z') = (\mathbf{1}_{q_0}, z_0)$. Hence $(f_n', z_n') \in \mathcal{V}$ for each large n and thus $(f_n', z_n') = (f_n, z_n)$. This is a contradiction since $[f_n', z_n']$ is a singular and $[f_n, z_n]$ is a regular point of ev_{J_n}.   □

Now pick a regular almost complex structure J_b in a neighbourhood of J_V and a point p_b in a neighbourhood of p_0 such that p_b is a regular value of ev_{J_b}.

The strategy in order to prove the existence of a rational J_V-curve passing through p_0 is now as follows. First, one shows the next

Claim. There exists a rational J_b-holomorphic curve representing A and passing through p_b.

The proof of this claim uses the compactness theorem (Corollary 4.2) and inverse function theorem arguments. The latter are the reason for picking regular values p_b and J_b. And since J_b can be chosen arbitrarily close to J_V and p_b arbitrarily close to p_0, the compactness theorem will again apply and provide a rational J_V-holomorphic curve passing through the point p_0.

Proof of the claim. Choose p_a and J_a in a neighbourhood of $(p_0, j \oplus J_M)$ according to Lemma 4.3. By the transversality theorem, there exists a smooth path $\alpha: [0, 1] \to \mathcal{J}$ connecting J_a to J_b and transverse to the projection $P: \mathcal{M}(A, \mathcal{J}) \to \mathcal{J}$. Consequently,

$$M(A, \alpha) := P^{-1}(\alpha[0, 1])$$

is a smooth manifold with boundary $\partial M(A, \alpha) = M(A, J_a) \sqcup M(A, J_b)$ of finite dimension. The Lie group G acts also on $M(A, \alpha) \times S^2$ as in (4.2). Observe that the points $(p_a, 0)$ and $(p_b, 1)$ are regular values for the evaluation map

$$\mathrm{ev}_\alpha: M(A, \alpha) \times_G S^2 \to V \times [0, 1]$$

mapping the equivalence class $[(f, z)]$ represented by an $\alpha(t)$-holomorphic curve f and $z \in S^2$ to $(f(z), t)$. Therefore, a smooth path $\beta: [0, 1] \to V \times [0, 1]$ can be chosen from $(p_a, 0)$ to $(p_b, 1)$ transverse to ev_α and transverse to the boundary $V \times \{0\} \sqcup V \times \{1\}$.

$M(A, \alpha[0, 1]) \times_G S^2$

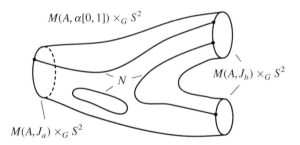

$M(A, J_b) \times_G S^2$

N

$M(A, J_a) \times_G S^2$

Figure 25.

Then $N := \mathrm{ev}_\alpha^{-1}(\beta[0, 1])$ is a smooth submanifold of $M(A, \alpha) \times_G S^2$ with boundary

$$\partial N = N \cap \partial(M(A, \alpha) \times_G S^2) = (N \cap M(A, J_a) \times_G S^2) \sqcup (N \cap M(A, J_b) \times_G S^2).$$

The goal is to find a J_b-holomorphic curve passing through p_b, that is to say, a point in $N \cap M(A, J_b) \times_G S^2$. To that end, we apply Corollary 4.2, which implies that $M(A, \alpha) \times_G S^2$ and hence N is compact (see Figure 25).

The piece $N \cap M(A, J_a) \times_G S^2$ of ∂N consists of exactly one point by Lemma 4.3. In particular, N is one-dimensional. As a compact one-dimensional manifold with bound-

ary, N has an even number of boundary points. Hence $N \cap M(A, J_b) \times_G S^2 \neq \emptyset$ and
this finishes the proof of the claim.   □

Remark. It should be stressed that in the previous proof, if $M(A, \alpha) \times_G S^2$ were not
compact, nothing could be said about $N \cap M(A, J_b) \times_G S^2$.

It is now easy to finish the proof of Proposition 2.1. The results proved so far im-
ply that there exists a sequence $(J_n, p_n)_{n \geq 1}$ in $\mathscr{J} \times V$ converging to (J_V, p_0) and a
sequence $(f_n)_{n \geq 1}$ of rational J_n-holomorphic curves in the homology class A such that
p_n is contained in the image of f_n. By Corollary 4.2, there is a sequence $\sigma_n \in G$ such
that $(f_n \circ \sigma_n)$ converges uniformly to a rational J_V-holomorphic curve f. Clearly, f
represents A and contains p in its image.

Appendix A

The classical isoperimetric inequality

For the convenience of the reader we shall give a proof of the classical isoperimetric inequality for maps of compact surfaces with boundary to the Euclidean space $\mathbb{R}^m$. The proof presented here is a special case of [BZ] p. 61f and p. 128f.

Proposition A.1. *Let S be a compact surface with boundary and $\gamma: \partial S \to \mathbb{R}^m$ a smooth curve. Then there exists a smooth map $g: S \to \mathbb{R}^m$ with $g|_{\partial S} = \gamma$ and*

$$4\pi \mathcal{A}(g) \leq \ell^2(\gamma).$$

Moreover, one can require that $g(S)$ is contained in the convex hull of $\gamma(\partial S)$.

Lemma A.2 (Wirtinger lemma). *Assume $\varphi: S^1 = 2\pi\mathbb{Z}\backslash\mathbb{R} \to \mathbb{R}^m$ is a smooth map and $\int_0^{2\pi} \varphi(\theta)\, d\theta = 0$. Then*

$$\int_0^{2\pi} \langle \dot\varphi, \dot\varphi \rangle\, d\theta \geq \int_0^{2\pi} \langle \varphi, \varphi \rangle\, d\theta.$$

Proof. It is sufficient to check this for $m = 1$. We consider the Fourier series of φ and $\dot\varphi$ with respect to the basis $e_n: S^1 \to \mathbb{C}$, $e_n(\theta) = (2\pi)^{-1/2} e^{in\theta}$, $n \in \mathbb{Z}$:

$$\varphi = \sum_{n\in\mathbb{Z}} a_n e_n \quad \text{and} \quad \dot\varphi = \sum_{n\in\mathbb{Z}} ina_n e_n$$

with suitable $a_n \in \mathbb{C}$. Thus

$$\int_0^{2\pi} \varphi^2\, d\theta = \sum_{n\in\mathbb{Z}} |a_n|^2 \quad \text{and} \quad \int_0^{2\pi} \dot\varphi^2\, d\theta = \sum_{n\in\mathbb{Z}} |ina_n|^2.$$

Note that $a_0 = 0$ since by assumption the mean value of φ is zero. This implies the claim. $\qquad\square$

Proof of Proposition A.1. Let us first assume $S = \overline{D} \subset \mathbb{C}$ is the closed unit disc. We may assume without loss of generality that γ has constant speed $\|\dot\gamma\| \equiv \ell(\gamma)/2\pi$. Then the centre of gravity of γ is $\ell(\gamma)^{-1} \int_0^{2\pi} \gamma(\theta)\, d\theta$ and we may assume that it is zero. We pick a smooth, non-decreasing function $\beta: \mathbb{R} \to \mathbb{R}$ satisfying

$$\beta(r) = \begin{cases} 0, & \text{if } r \leq 1/3 \\ 1, & \text{if } r \geq 2/3, \end{cases}$$

and define $g \colon \overline{D} \to \mathbb{R}^m$ by $g(z) = \beta(|z|) \cdot \gamma(z/|z|)$. We claim that g has the required properties. For $t > 0$ we put $g_t := t \cdot g$ and we observe that $\mathcal{A}(g_t) = t^2 \mathcal{A}(g)$. This implies that

$$2\mathcal{A}(g) = \frac{d\mathcal{A}(g_t)}{dt}\bigg|_{t=1} = \int_0^{2\pi} \left(\|\gamma\|^2 \|\dot{\gamma}\|^2 - \langle \gamma, \dot{\gamma} \rangle^2 \right)^{1/2} d\theta$$

$$\leq \int_0^{2\pi} \|\gamma\| \, \|\dot{\gamma}\| \, d\theta$$

$$\leq \left(\int_0^{2\pi} \|\gamma\|^2 \, d\theta \right)^{1/2} \left(\int_0^{2\pi} \|\dot{\gamma}\|^2 \, d\theta \right)^{1/2}$$

$$\leq \int_0^{2\pi} \|\dot{\gamma}\|^2 \, d\theta = \frac{\ell^2(\gamma)}{2\pi},$$

where the last inequality follows from Lemma A.2. The equality in the first line can be easily deduced by observing that $(0, t) \times S^1 \ni (r, \theta) \mapsto r \cdot \gamma(\theta)$ is a reparametrization of $g_t|_{\overline{D} \setminus \{0\}}$. This concludes the proof in case $S = D$.

For arbitrary S we choose a closed collar $\mathscr{C}$ around ∂S in S and identify $\mathscr{C} \simeq [-1, 1] \times \partial S$ in such a way that $\partial S \simeq \{1\} \times \partial S$. Let $\partial_1 S, \ldots, \partial_k S$ denote the boundary components of S. Each component is diffeomorphic to S^1. Let x_ν be the centre of gravity of $\gamma|_{\partial_\nu S}$. Then the map $g \colon [-1, 1] \times \partial S \to \mathbb{R}^m$ given by

$$g(r, s) := \begin{cases} x_\nu + \beta(r)(\gamma(s) - x_\nu), & \text{if } (r, s) \in [0, 1] \times \partial_\nu S \\ x_\nu + \beta(-r)(x_1 - x_\nu), & \text{if } (r, s) \in [-1, 0] \times \partial_\nu S \end{cases}$$

extends to a smooth map $g \colon S \to \mathbb{R}^m$ by $g(z) = x_1$ for $z \in S \setminus \mathscr{C}$. This map g has the desired properties, namely

$$4\pi \mathcal{A}(g) = 4\pi \sum_{\nu=1}^k \mathcal{A}(g|_{[0,1] \times \partial_\nu S}) \leq \sum_{\nu=1}^k \ell^2(\gamma|_{\partial_\nu S}) \leq \ell^2(\gamma).$$

We note that $g(S)$ is contained in the convex hull of $\gamma(\partial S)$, and this finishes the proof.

$\square$

Appendix B

The C^k-topology

In this appendix we briefly introduce the notion of C^k-convergence for C^k-maps between manifolds. The reader can find a comprehensive description in [Hi].

First let us fix some notation. By M, N we denote $(C^\infty$-)manifolds of dimensions m, n, respectively. Charts are denoted by (φ, U), where φ is the map and U its domain. By $\mathscr{L}^r(\mathbb{R}^m, \mathbb{R}^n)$ we mean the space of r-linear maps

$$\underbrace{\mathbb{R}^m \times \cdots \times \mathbb{R}^m}_{r \text{ times}} \to \mathbb{R}^n,$$

and $\| \ \|$ denotes the standard norm on the corresponding vector space.

Let $V \subset \mathbb{R}^m$ and $W \subset \mathbb{R}^n$ be open subsets. For a C^k-map $f: V \to W$, k a non-negative integer, we denote by $f^{(k)}$ its k-jet, that is

$$f^{(k)}: V \to W \times \mathscr{L}^1(\mathbb{R}^m, \mathbb{R}^n) \times \cdots \times \mathscr{L}^k(\mathbb{R}^m, \mathbb{R}^n),$$
$$f^{(k)}(x) = \left(f(x), Df(x), \ldots, D^k f(x) \right).$$

Definition. Let k be a non-negative integer. A sequence $f_v: V \to W$, $v \geq 1$, of C^k-maps *converges in the C^k-topology* to a C^k-map $f: V \to W$ if the sequence $(f_v^{(k)})$ of k-jets converges locally uniformly to $f^{(k)}$. Then (f_v) is also said to *converge in C^k*. In other words, (f_v) converges in C^k if all derivatives from order 0 up to order k converge locally uniformly.

A sequence $f_v: V \to W$, $v \geq 1$, of C^∞-maps *converges in C^∞* if it converges in each C^k-topology, $0 \leq k < \infty$.

Remark. C^k-topology, in the sense defined here, is sometimes, for example in [Hi], called the weak C^k-topology or the topology of local C^k-convergence.

Lemma B.1. *Let $U \subset \mathbb{R}^l$, $V \subset \mathbb{R}^m$ and $W \subset \mathbb{R}^n$ be open subsets. Assume $g_v: U \to V$ and $f_v: V \to W$, $v \geq 1$, are two sequences of C^k-maps converging in C^k to g and f, respectively. Then $(f_v \circ g_v)$ converges in C^k to $f \circ g$.*

Proof. One estimates the derivatives of order $0 \leq r \leq k$ of the difference

$$f_v \circ g_v - f \circ g = f_v \circ g_v - f \circ g_v + f \circ g_v - f \circ g$$

as follows:

$$\| D^r(f_v \circ g_v - f \circ g) \| \leq \| D^r(f_v \circ g_v - f \circ g_v) \| + \| D^r(f \circ g_v - f \circ g) \|.$$

The components of $D^r(f_\nu \circ g_\nu - f \circ g_\nu)$ and $D^r(f \circ g_\nu - f \circ g)$, respectively, i.e., differences of products of partial derivatives, converge locally uniformly to zero since (f_ν) and (g_ν) converge in C^k.   □

Lemma B.2. *Let $k \geq 1$ be some positive integer and $f_\nu: U \to V$ be a sequence of diffeomorphisms between open subsets of $\mathbb{R}^m$. If (f_ν) converges in C^k to a diffeomorphism f, then the inverse maps (f_ν^{-1}) converge in C^k to f^{-1}.*

Proof. First of all we show that the sequence (f_ν^{-1}) converges in C^0 to f^{-1}. This is equivalent to the following

Claim. Given any $r > 0$, any $y_0 \in V$ and any open subset $U' \subset U$ such that the closed Euclidean ball $\overline{B}_r(y_0) \subset \mathbb{R}^m$ is contained in V and $f^{-1}(\overline{B}_r(y_0)) \subset U'$, then $f_\nu^{-1}(\overline{B}_r(y_0)) \subset U'$ provided ν is sufficiently large.

Without loss of generality we may assume that $U = V$ and $f = \mathrm{id}$. Under the assumption in the claim we choose some $\varepsilon > 0$ such that $\overline{B}_{r+\varepsilon}(y_0) = \mathrm{id}^{-1}(\overline{B}_{r+\varepsilon}(y_0)) \subset U'$. Then for any sufficiently large ν and each $y \in \overline{B}_r(y_0)$ the map

$$F_{\nu,y}: x \mapsto (\mathrm{id} - f_\nu)(x) + y$$

is a distance shrinking map

$$F_{\nu,y}: \overline{B}_{r+\varepsilon}(y_0) \to \overline{B}_{r+\varepsilon}(y_0)$$

by the mean value theorem since (f_ν) converges in C^1 to the identity. The fixed point x' of $F_{\nu,y}$ satisfies $f_\nu(x') = y$ and this proves the claim.

Now the C^k-convergence of (f_ν^{-1}) follows by induction in view of the commutative diagram

To that end observe that the inverse map $GL(m, \mathbb{R}) \to GL(m, \mathbb{R})$ is of class C^∞.   □

Definition. Let k be a non-negative integer or $k = \infty$ and $f_\nu: M \to N$, $\nu \geq 1$, a sequence of C^k-maps. The sequence (f_ν) *converges in the C^k-topology* to a C^k-map f if for each $p \in M$ there exists a chart (φ, U) near p and a chart (ψ, V) of N near $f(p)$ with $f(U) \subset V$ such that $\psi \circ f_\nu \circ \varphi^{-1}: \varphi(U) \to \psi(V)$ is well defined for each sufficiently large ν and converges in C^k to $\psi \circ f \circ \varphi^{-1}$. By Lemma B.1 this definition is independent of the choice of the charts (φ, U) and (ψ, V).

Remarks. 1. Obviously Lemma B.1 and B.2 generalize to maps between manifolds.

2. A sequence of Riemannian metrics $(g_v)_{v \geq 1}$ and a sequence of almost complex structures $(J_v)_{v \geq 1}$ on a manifold M *converge in C^k* if for any vector fields X, Y on M the functions $g_v(X, Y)$ and the vector fields $J_v X$, respectively, converge in C^k as $v \to \infty$. Of course, it suffices to check this for local coordinate fields where the coordinates vary in some atlas.

Proposition B.3. *Let $k \geq 0$ be an integer. Assume $f_v \colon M \to N$, $v \geq 1$, is a sequence of C^{k+1}-maps converging in C^0 to a continuous map $f \colon M \to N$. Assume that for each $p \in M$ there exist charts (φ, U) near p and (ψ, V) near $f(p)$ such that the following holds. There exists a neighbourhood $W \subset \varphi(U)$ of $\varphi(p)$ and a constant c such that*

$$\|D^r(\psi \circ f_v \circ \varphi^{-1})(x)\| \leq c \qquad \textbf{(B.1)}$$

for each $x \in W$, each sufficiently large v and each $r = 1, \ldots, k+1$. Then f is of class C^k and (f_v) converges even in C^k to f.

Remark B.4. Since (f_v) converges in the C^0-topology, (B.1) makes sense. The proposition says that a C^0-convergent sequence whose derivatives up to order $k+1$ are locally bounded is indeed C^k-convergent.

Proof. Without loss of generality we may assume that $M \subset \mathbb{R}^m$ is open, $N = \mathbb{R}^n$, $(\varphi, U) = (\text{id}_M, M)$ and $(\psi, V) = (\text{id}_{\mathbb{R}^n}, \mathbb{R}^n)$. For (f_v) we consider the sequence $(f_v^{(k)})$ of k-jets. By assumption (B.1) on each compact subset of U the sequence $(\|Df_v^{(k)}\|)_{v \geq 1}$ is uniformly bounded. Thus for each $x \in U$ the set $\{ f_v^{(k)}(x) \mid v \geq 1 \}$ is relatively compact in $\mathbb{R}^n \times \cdots \times \mathscr{L}^k(\mathbb{R}^m, \mathbb{R}^n)$ since additionally (f_v) converges in C^0. Applying Arzelà-Ascoli's theorem some subsequence $(f_{v_i}^{(k)})$ converges in C^0 to a continuous map

$$g \colon U \to \mathbb{R}^n \times \cdots \times \mathscr{L}^k(\mathbb{R}^m, \mathbb{R}^n).$$

We denote by $g^0 \colon U \to \mathbb{R}^n$ and $g^r \colon U \to \mathscr{L}^r(\mathbb{R}^m, \mathbb{R}^n)$ the components of g. Then $g^0 = f$ and the sequence $(D^r f_{v_i})_{i \geq 1}$ converges in C^0 to g^r. By induction one easily shows that f is of class C^k and $D^r f = g^r$ for $r = 1, \ldots, k$. Hence (f_{v_i}) converges in C^k to f.

If (f_v) did not converge in C^k to f one could find a C^0-convergent subsequence of $(f_v^{(k)})$ not converging to $g = f^{(k)}$ by Arzelà-Ascoli's theorem. This would contradict the C^0-convergence of (f_v) to f. $\qquad\square$

References on pseudo-holomorphic curves

We present here some references related to pseudo-holomorphic curves and their applications in symplectic geometry, contact geometry, Floer homology, quantum cohomology and in the theory of smooth 4-manifolds. It should be stressed that the following list is by no means complete.

Important tools for studying the space of solutions for the generalized Cauchy-Riemann-equations are Fredholm theory and methods from global analysis (see for example, [2], [22], [29] or [39]). Recently in [41] and [42], pseudo-holomorphic curves in symplectic 4-manifolds were constructed from solutions of the Seiberg-Witten equations.

For further studies we refer the reader to the textbooks [1], [2], [3] and [29].

[1]   C. ABBAS & H. HOFER, *Holomorphic curves and global questions in contact geometry*, to be published by Birkhäuser, Basel · Boston · Berlin.

[2]   B. AEBISCHER & M. BORER & M. KÄLIN & CH. LEUENBERGER & H.M. REIMANN, *Symplectic Geometry*, An Introduction based on the Seminar in Berne 1992, Progress in Mathematics **124**, Birkhäuser, Basel · Boston · Berlin, 1994.

[3]   M. AUDIN & J. LAFONTAINE (editors), *Holomorphic curves in symplectic geometry*, Progress in Mathematics **117**, Birkhäuser, Basel · Boston · Berlin, 1994.

[4]   M. BETZ & J. RADE, Products and relations in symplectic Floer homology, *preprint*, 1995.

[5]   Y. ELIASHBERG, Filling by holomorphic discs and its applications, *London Mathematical Society Lecture Note Series* **151** (1991), 45–67.

[6]   Y. ELIASHBERG & L. POLTEROVICH, Unknottedness of Lagrangian surfaces in symplectic 4-manifolds, *International Mathematics Research Notices* **11** (1993), 295–301.

[7]   A. FLOER, The unregularized gradient flow of the symplectic action, *Communications on Pure and Applied Mathematics* **41** (1988), 775–813.

[8]   A. FLOER, Morse theory for Lagrangian intersections, *Journal of Differential Geometry* **28** (1988), 513–547.

[9]   A. FLOER, Symplectic fixed points and holomorphic spheres, *Communications in Mathematical Physics* **120** (1989), 207–221.

[10]   A. GIVENTAL & B. KIM, Quantum Cohomology of flag manifolds and Toda lattices, *Communications in Mathematical Physics* **168** (1995), 609–641.

[11]   M. GROMOV, Pseudo holomorphic curves in symplectic manifolds, *Inventiones mathematicae* **82** (1985), 307–347.

[12]   H. HOFER, Pseudoholomorphic curves in symplectizations with applications to the Weinstein conjecture in dimension three, *Inventiones mathematicae* **114** (1993), 515–563.

[13]   H. HOFER & V. LIZAN & J.-C. SIKORAV, On genericity for holomorphic curves in 4-dimensional almost-complex manifolds, *preprint*, 1994.

[14]   H. HOFER & C. TAUBES & A. WEINSTEIN & E. ZEHNDER (editors), *The Floer Memorial Volume*, Progress in Mathematics **133**, Birkhäuser, Basel · Boston · Berlin, 1995.

[15]   H. HOFER & C. VITERBO, The Weinstein conjecture in the presence of holomorphic spheres, *Communications on Pure and Applied Mathematics* **45** (1992), 583–622.

[16]   H. HOFER & K. WYSOCKI & E. ZEHNDER, Properties of pseudo-holomorphic curves in symplectisations I, to appear in *Annales de l'Institut Henri Poincaré, Analyse non linéaire*.

[17]   H. HOFER & K. WYSOCKI & E. ZEHNDER, Properties of pseudo-holomorphic curves in symplectisations II: Embedding Controls and Algebraic Invariants, *GAFA Geometric And Functional Analysis* **5** (1995), 270–328.

[18]   H. HOFER & K. WYSOCKI & E. ZEHNDER, Properties of pseudo-holomorphic curves in symplectisations III: Fredholm theory, *preprint*, 1996.

[19]   M. KONTSEVICH & Y. MANIN, Gromov-Witten classes, quantum cohomology and enumerative geometry, *Communications in Mathematical Physics* **164** (1994), 525–562.

[20]   F. LABOURIE, Immersions isométriques elliptiques et courbes pseudo-holomorphes, *Journal of Differential Geometry* **30** (1989), 395–424.

[21]   F. LAUDENBACH, Orbites périodiques et courbes pseudo-holomorphes, application à la conjecture de Weinstein en dimension 3 [d'après H. Hofer et al.], *Séminaire Bourbaki*, exposé **786** (1993–1994).

[22]   D. MCDUFF, Examples of symplectic structures, *Inventiones mathematicae* **89** (1987), 13–36.

[23]   D. MCDUFF, The structure of rational and ruled symplectic 4-manifolds, *Journal of the American Mathematical Society* **3** (1990), 679–712.

[24]   D. MCDUFF, Elliptic methods in symplectic geometry, *Bulletin of the American Mathematical Society* **23** (1990), 311–358.

[25]   D. MCDUFF, The local behaviour of J-holomorphic curves in almost complex 4-manifolds, *Journal of Differential Geometry* **34** (1991), 143–164.

[26]   D. MCDUFF, Immersed spheres in symplectic 4-manifolds, *Annales de l'Institut Fourier* **42** (1992), 369–392.

[27]   D. MCDUFF, Singularities of J-holomorphic curves in almost complex 4-manifolds, *Journal of Geometric Analysis* **3** (1992), 149–266.

[28]   D. MCDUFF, J-holomorphic spheres in symplectic 4-manifolds: a survey, *preprint*, Stony Brook, 1995.

[29]   D. MCDUFF & D. SALAMON, *J-holomorphic curves and quantum cohomology*, University lecture series, vol. **6**, American Mathematical Society, Providence, RI, 1994.

[30]   M. MICALLEF & B. WHITE, The structure of branch points in area minimizing surfaces and in pseudo-holomorphic curves, *Annals of Mathematics* **141** (1995), 35–85.

[31]   J. MOSER, On the persistence of pseudo-holomorphic curves on an almost complex torus (with an appendix by Jürgen Pöschel), *Inventiones mathematicae* **119** (1995), 401–442.

[32] Y.-G. OH, Removal of boundary singularities of pseudoholomorphic curves with Lagrangian boundary conditions, *Communications on Pure and Applied Mathematics* **45** (1992), 121–139.

[33] T. PARKER & J. WOLFSON, Pseudoholomorphic maps and bubble trees, *Journal of Geometric Analysis* **3** (1993), 63–98.

[34] Y. RUAN, Virtual neighborhoods and pseudo-holomorphic curves, *preprint*, 1996.

[35] Y. RUAN & G. TIAN, Bott-type symplectic Floer cohomology and its multiplication structures, *Mathematical Research Letters* **2** (1995), 203–219.

[36] Y. RUAN & G. TIAN, A mathematical theory of quantum cohomology, *Journal of Differential Geometry*, **42** (1995), 259–367.

[37] Y. RUAN & G. TIAN, Higher genus symplectic invariants and sigma model coupled with gravity, *Turkish Journal of Mathematics* **20** (1996), 75-83.

[38] D. SALAMON (editor), *Symplectic Geometry*, London Mathematical Society Lecture Note Series **192**, Cambridge University Press, 1993.

[39] M. SCHWARZ, Cohomology Operations from S^1-Cobordisms in Floer Homology, *thesis*, Diss. ETH No. 11182, Zürich, 1995.

[40] J. - C. SIKORAV, Singularities of J-Holomorphic Curves, *preprint*, 1995.

[41] C. H. TAUBES, The Seiberg-Witten and Gromov Invariants, *Mathematical Research Letters* **2** (1995), 221–238.

[42] C. H. TAUBES, SW $\Rightarrow$ Gr: From Seiberg-Witten equations to pseudo-holomorphic curves, *Journal of the American Mathematical Society* **9** (1996), 845–918.

[43] C. B. THOMAS (editor), *Contact and Symplectic Geometry*, Publications of the Newton Institute, Cambridge University Press, 1996.

[44] R. YE, Gromov's compactness theorem for pseudo holomorphic curves, *Transactions of the American Mathematical Society* **342** (1994), 671–694.

Bibliography

[Ab]   W. ABIKOFF, *The Real Analytic Theory of Teichmüller Space*, Lecture Notes in Mathematics **820**, Springer-Verlag, Berlin Heidelberg New York, 1980.

[ABK]   B. AEBISCHER & M. BORER & M. KÄLIN & CH. LEUENBERGER & H.M. REIMANN, *Symplectic Geometry*, An Introduction based on the Seminar in Berne 1992, Progress in Mathematics **124**, Birkhäuser, Basel · Boston · Berlin, 1994.

[AK]   D. V. ANOSOV & A.B. KATOK, New examples in smooth ergodic theory. Ergodic diffeomorphisms, *Transactions of the Moscow Mathematical Society* **23** (1970), 1–35.

[Ad]   M. AUDIN, Symplectic and almost complex manifolds, in [AL].

[AL]   M. AUDIN & J. LAFONTAINE (editors), *Holomorphic curves in symplectic geometry*, Progress in Mathematics **117**, Birkhäuser, Basel · Boston · Berlin, 1994.

[Ba]   A. BANYAGA, Formes-volume sur les variétés a bord, *L'Enseignement mathématique* **20** (1974), 127–131.

[BP]   R. BENEDETTI & C. PETRONIO, *Lectures on Hyperbolic Geometry*, Springer-Verlag, Berlin Heidelberg New York, 1992.

[Bu]   P. BUSER, *Geometry and Spectra of Compact Riemann Surfaces*, Progress in Mathematics **106**, Birkhäuser, Basel · Boston · Berlin, 1992.

[BZ]   Y. BURAGO & V. ZALGALLER, *Geometric Inequalities*, Grundlehren der mathematischen Wissenschaften **285**, Springer-Verlag, Berlin Heidelberg New York, 1988.

[Ch]   I. CHAVEL, *Riemannian geometry: a modern introduction*, Cambridge University Press, 1993.

[dC]   M. P. DO CARMO, *Riemannian Geometry*, Birkhäuser, Basel · Boston · Berlin, 1992.

[DM]   P. DELIGNE & D. MUMFORD, The irreducibility of the space of curves of given genus, *Publications Mathématiques Institute des Hautes Études Scientifiques*, **36** (1969), 75–109.

[EH]   I. EKELAND & H. HOFER, Symplectic topology and Hamiltonian dynamics, *Mathematische Zeitschrift*, **200** (1989), 355–378.

[Fe]   H. FEDERER, *Geometric Measure Theory*, Grundlehren der mathematischen Wissenschaften **153**, Springer-Verlag, Berlin Heidelberg New York, 1969.

[FK]   H. FARKAS & I. KRA, *Riemann Surfaces*, Graduate Texts in Mathematics **71**, Springer-Verlag, New York Heidelberg Berlin, 1980.

[FLP]   A. FATHI & F. LAUDENBACH & V. POÉNARU, *Traveaux de Thurston sur les surfaces*, Astérisque **66–67**, Société Mathématique de France, Paris, 1979.

[GHL]   S. GALLOT & D. HULIN & J. LAFONTAINE, *Riemannian Geometry*, second edition, Springer-Verlag, Berlin Heidelberg New York, 1990.

[Go]   R. GOMPF, A new construction of symplectic manifolds, *Annals of Mathematics* **142** (1995), 527–595.

[Gr]   M. GROMOV, Pseudo holomorphic curves in symplectic manifolds, *Inventiones mathematicae* **82** (1985), 307–347.

[Hi]   M. HIRSCH, *Differential Topology*, Graduate Texts in Mathematics **33**, Springer-Verlag, New York Berlin Heidelberg, 1976.

[HZ]   H. HOFER & E. ZEHNDER, *Symplectic Invariants and Hamiltonian Dynamics*, Birkhäuser Advanced Text, Birkhäuser, Basel · Boston · Berlin, 1994.

[Jo]   J. JOST, *Riemannian Geometry and Geometric Analysis*, Springer-Verlag, Berlin Heidelberg New York, 1995.

[Kl]   W. KLINGENBERG, *A course in differential geometry*, Springer-Verlag, Berlin Heidelberg New York, 1978.

[Lf]   J. LAFONTAINE, Some relevant Riemannian geometry, in [AL].

[LM]   F. LALONDE & D. MCDUFF, The geometry of symplectic energy, *Annals of Mathematics* **141** (1995), 349–371.

[Lg1]   S. LANG, *Real and Functional Analysis*, third edition, Graduate Texts in Mathematics **142**, Springer-Verlag, New York Berlin Heidelberg, 1993.

[Lg2]   S. LANG, *Differential manifolds*, Springer-Verlag, New York Berlin Heidelberg, 1985.

[Ma]   W. S. MASSEY, *A Basic Course in Algebraic Topology*, Graduate Texts in Mathematics **127**, Springer-Verlag, New York Berlin Heidelberg, 1991.

[MS1]   D. MCDUFF & D. SALAMON, *Introduction to Symplectic topology*, Oxford Mathematical Monographs, Oxford University Press, 1995.

[MS2]   D. MCDUFF & D. SALAMON, *J-holomorphic curves and quantum cohomology*, University lecture series, vol. **6**, American Mathematical Society, Providence, RI, 1994.

[Mi]   J. W. MILNOR, *Topology from the Differentiable Viewpoint*, The University Press of Virginia, Charlottesville, 1965.

[MiS]   J. W. MILNOR & J.D. STASHEFF, *Characteristic Classes*, Annals of Mathematics Studies **76**, Princeton University Press and University of Tokyo Press, Princeton, New Jersey, 1974.

[Mo]   E. E. MOISE, *Geometric Topology in Dimension 2 and 3*, Graduate Texts in Mathematics **47**, Springer-Verlag, New York Berlin Heidelberg, 1977.

[Ms]   J. MOSER, On volume elements on a manifold, *Transactions of the American Mathematical Society* **120** (1965), 280–296.

[Mu]   M.-P. MULLER, Gromov's Schwarz lemma as an estimate of the gradient for holomorphic curves, in [AL].

[NN]   A. NEWLANDER & L. NIRENBERG, Complex analytic coordinates in almost complex manifolds, *Annals of Mathematics* **65** (1957), 391–404.

[Pa1]   P. PANSU, Notes sur les pages 316 à 323 de l'article de M. Gromov: Pseudoholomorphic curves in symplectic manifolds, *preprint*, Ecole Polytechnique, Palaiseau, 1986.

[Pa2]   P. PANSU, Compactness, in [AL].

[PW]   T. PARKER & J. WOLFSON, Pseudoholomorphic maps and bubble trees, *Journal of Geometric Analysis* **3** (1993), 63–98.

[SaU]   J. SACKS & K. UHLENBECK, The existence of minimal immersions of 2-spheres, *Annals of Mathematics* **113** (1981), 1–24.

[Si]   J. - C. SIKORAV, Some properties of holomorphic curves in almost complex manifolds, in [AL].

[Sm]   S. SMALE, An infinite dimensional version of Sard's theorem, *American Journal of Mathematics* **87** (1965), 861–866.

[Sp]   M. SPIVAK, *A Comprehensive Introduction to Differential Geometry*, Vol. IV, second edition, Publish or Perish Inc., Houston, Texas, 1979, pp. 455ff.

[St]   E. STEIN, *Singular Integrals and Differentiability Properties of Functions*, Princeton Mathematical Series **30**, Princeton University Press, Princeton, New Jersey, 1970.

[Tr]   A. TROMBA, *Teichmüller Theory in Riemannian Geometry*, Lectures in Mathematics ETH Zürich, Birkhäuser, Basel · Boston · Berlin, 1992.

[Wh]   H. WHITNEY, Analytic extensions of differentiable functions defined in closed sets, *Transactions of the American Mathematical Society* **36** (1934), 63–89.

[Ye]   R. YE, Gromov's compactness theorem for pseudo holomorphic curves, *Transactions of the American Mathematical Society* **342** (1994), 671–694.

Index

Progress in Mathematics

Edited by:

H. Bass
Columbia University
New York
10027
U.S.A.

J. Oesterlé
Dépt. de Mathématiques
Université de Paris VI
4, Place Jussieu
75230 Paris Cedex 05, France

A. Weinstein
Dept. of Mathematics
University of CaliforniaNY
Berkeley, CA 94720
U.S.A.

Progress in Mathematics is a series of books intended for professional mathematicians and scientists, encompassing all areas of pure mathematics. This distinguished series, which began in 1979, includes authored monographs, and edited collections of papers on important research developments as well as expositions of particular subject areas.

We encourage preparation of manuscripts in such form of TeX for delivery in camera-ready copy which leads to rapid publication, or in electronic form for interfacing with laser printers or typesetters.

Proposals should be sent directly to the editors or to: Birkhäuser Boston, 675 Massachusetts Avenue, Cambridge, MA 02139, U.S.A.

BAT • Birkhäuser Advanced Texts
Basler Lehrbücher

H. Hofer / E. Zehnder, ETH, Mathematik, Zürich, Switzerland

Symplectic Invariants and Hamiltonian Dynamics

1994. 342 pages. Hardcover
ISBN 3-7643-5066-0

The discoveries of the last decade have opened new perspectives on the old field of Hamiltonian systems and have led to the creation of a new field: symplectic topology. Surprising rigidity phenomena demonstrate that the nature of symplectic mappings is very different from that of volume preserving mappings. This raises new questions, many of them still unanswered. On the other hand, analysis of an old variational principle in classical mechanics has established global periodic phenomena in Hamiltonian systems. As it turns out, these seemingly different phenomena are mysteriously related. One of the links is a class of symplectic invariants, called symplectic capacities. These invariants are the main theme of this book, which includes such topics as basic symplectic geometry, symplectic capacities and rigidity, periodic orbits for Hamiltonian systems and the least action principle, a bi-invariant metric on the symplectic diffeomorphism group and its geometry, symplectic fixed point theory, the Arnold conjectures and first order elliptic systems, and a survey on Floer homology and symplectic homology.

"Symplectic Topology has become a fascinating subject of research over the past fifteen years... This book is written by two experienced researchers, will certainly fill in a gap in the theory of symplectic topology. The authors have taken part in the development of such a theory by themselves or by their collaboration with other outstanding people in the area... All the chapters have a nice introduction with the historic developement of the subject and with a perfect description of the state of the art."

ZENTRALBLATT MATHEMATIK, 1995

"This book is a beautiful introduction to one outlook on the exciting new developments of the last ten to fifteen years in symplectic geometry, or symplectic topology..."

MATHEMATICAL REVIEWS, 96g 1996

For orders originating from all over
the world except USA and Canada:
Birkhäuser Verlag AG
P.O Box 133
CH-4010 Basel/Switzerland
Fax: +41/61/205 07 92
e-mail: farnik@birkhauser.ch

For orders originating in the
USA and Canada:
Birkhäuser
333 Meadowland Parkway
USA-Secaurus, NJ 07094-2491
Fax: +1 201 348 4033
e-mail: orders@birkhauser.com

Birkhäuser

Birkhäuser Verlag AG
Basel · Boston · Berlin

VISIT OUR HOMEPAGE **http://www.birkhauser.ch**

PM 117 • Progress in Mathematics

M. Audin, University Louis Pasteur, IRMA, Strasbourg, France / **J. Lafontaine**, University Montpellier 2, Montpellier, France (Eds)

Holomorphic curves in symplectic geometry

1994. 340 pages. Hardcover
ISBN 3-7643-2997-1

This book is devoted to pseudo-holomorphic curve methods in symplectic geometry. It contains an introduction to symplectic geometry and relevant techniques of Riemannian geometry, proofs of Gromov's compactness theorem, an investigation of local properties of holomorphic curves, including positivity of intersections, and applications to Lagrangian embeddings problems. The chapters are based on a series of lectures given previously by *M. Audin, A. Banyaga, P. Gauduchon, F. Labourie, J. Lafontaine, F. Lalonde, Gang Liu, D. McDuff, M.-P. Muller, P. Pansu, L. Polterovich, J.C. Sikorav.*
In an attempt to make this book accessible also to graduate students, the authors provide the necessary examples and techniques needed to understand the applications of the theory. The exposition is essentially self-contained and includes numerous exercises.

PM 144 • Progress in Mathematics

A. Bellaïche, Université de Paris VII-Denis Diderot, Paris, France / **J.-J. Risler**, Université Pierre et Marie Curie, Paris, France (Eds)

Sub-Riemannian Geometry

1996. 396 pages. Hardcover
ISBN 3-7643-5476-3

Sub-Riemannian geometry (also known as Carnot geometry in France, and non-holonomic Riemannian geometry in Russia) has been a full research domain for fifteen years, with motivations and ramifications in several areas of pure and applied mathematics, including *control theory • classical mechanics • Riemannian geometry (of which sub-Riemannian geometry constitutes a natural generalization, and where sub-Riemannian metrics may appear as limit cases) • diffusion on manifolds • analysis of hypoelliptic operators • Cauchy-Riemann (or CR) geometry*. Although links between these domains had been foreseen by many authors in the past, it is only in recent years that sub-Riemannian geometry has been recognized as a possible common framework for all these topics.
This book provides an introduction to sub-Riemannian geometry and presents the state of the art and open problems in the field. It consists of five coherent and original articles by the leading specialists.

For orders originating from all over the world except USA and Canada:
Birkhäuser Verlag AG
P.O Box 133
CH-4010 Basel/Switzerland
Fax: +41/61/205 07 92
e-mail: farnik@birkhauser.ch

For orders originating in the USA and Canada:
Birkhäuser
333 Meadowland Parkway
USA-Secaurus, NJ 07094-2491
Fax: +1 201 348 4033
e-mail: orders@birkhauser.com

Birkhäuser
Birkhäuser Verlag AG
Basel · Boston · Berlin

VISIT OUR HOMEPAGE **http://www.birkhauser.ch**